GUIDANCE FOR REVIEWING
SITE CHARACTERISTICS IN
SAFETY ANALYSIS REPORTS
FOR NUCLEAR INSTALLATIONS

The following States are Members of the International Atomic Energy Agency:

AFGHANISTAN	GEORGIA	OMAN
ALBANIA	GERMANY	PAKISTAN
ALGERIA	GHANA	PALAU
ANGOLA	GREECE	PANAMA
ANTIGUA AND BARBUDA	GRENADA	PAPUA NEW GUINEA
ARGENTINA	GUATEMALA	PARAGUAY
ARMENIA	GUINEA	PERU
AUSTRALIA	GUYANA	PHILIPPINES
AUSTRIA	HAITI	POLAND
AZERBAIJAN	HOLY SEE	PORTUGAL
BAHAMAS, THE	HONDURAS	QATAR
BAHRAIN	HUNGARY	REPUBLIC OF MOLDOVA
BANGLADESH	ICELAND	ROMANIA
BARBADOS	INDIA	RUSSIAN FEDERATION
BELARUS	INDONESIA	RWANDA
BELGIUM	IRAN, ISLAMIC REPUBLIC OF	SAINT KITTS AND NEVIS
BELIZE	IRAQ	SAINT LUCIA
BENIN	IRELAND	SAINT VINCENT AND
BOLIVIA, PLURINATIONAL	ISRAEL	THE GRENADINES
STATE OF	ITALY	SAMOA
BOSNIA AND HERZEGOVINA	JAMAICA	SAN MARINO
BOTSWANA	JAPAN	SAUDI ARABIA
BRAZIL	JORDAN	SENEGAL
BRUNEI DARUSSALAM	KAZAKHSTAN	SERBIA
BULGARIA	KENYA	SEYCHELLES
BURKINA FASO	KOREA, REPUBLIC OF	SIERRA LEONE
BURUNDI	KUWAIT	SINGAPORE
CABO VERDE	KYRGYZSTAN	SLOVAKIA
CAMBODIA	LAO PEOPLE'S DEMOCRATIC	SLOVENIA
CAMEROON	REPUBLIC	SOMALIA
CANADA	LATVIA	SOUTH AFRICA
CENTRAL AFRICAN	LEBANON	SPAIN
REPUBLIC	LESOTHO	SRI LANKA
CHAD	LIBERIA	SUDAN
CHILE	LIBYA	SWEDEN
CHINA	LIECHTENSTEIN	SWITZERLAND
COLOMBIA	LITHUANIA	SYRIAN ARAB REPUBLIC
COMOROS	LUXEMBOURG	TAJIKISTAN
CONGO	MADAGASCAR	THAILAND
COOK ISLANDS	MALAWI	TOGO
COSTA RICA	MALAYSIA	TONGA
CÔTE D'IVOIRE	MALDIVES	TRINIDAD AND TOBAGO
CROATIA	MALI	TUNISIA
CUBA	MALTA	TÜRKİYE
CYPRUS	MARSHALL ISLANDS	TURKMENISTAN
CZECH REPUBLIC	MAURITANIA	UGANDA
DEMOCRATIC REPUBLIC	MAURITIUS	UKRAINE
OF THE CONGO	MEXICO	UNITED ARAB EMIRATES
DENMARK	MONACO	UNITED KINGDOM OF
DJIBOUTI	MONGOLIA	GREAT BRITAIN AND
DOMINICA	MONTENEGRO	NORTHERN IRELAND
DOMINICAN REPUBLIC	MOROCCO	UNITED REPUBLIC OF TANZANIA
ECUADOR	MOZAMBIQUE	UNITED STATES OF AMERICA
EGYPT	MYANMAR	URUGUAY
EL SALVADOR	NAMIBIA	UZBEKISTAN
ERITREA	NEPAL	VANUATU
ESTONIA	NETHERLANDS,	VENEZUELA, BOLIVARIAN
ESWATINI	KINGDOM OF THE	REPUBLIC OF
ETHIOPIA	NEW ZEALAND	VIET NAM
FIJI	NICARAGUA	YEMEN
FINLAND	NIGER	ZAMBIA
FRANCE	NIGERIA	ZIMBABWE
GABON	NORTH MACEDONIA	
GAMBIA, THE	NORWAY	

The Agency's Statute was approved on 23 October 1956 by the Conference on the Statute of the IAEA held at United Nations Headquarters, New York; it entered into force on 29 July 1957. The Headquarters of the Agency are situated in Vienna. Its principal objective is "to accelerate and enlarge the contribution of atomic energy to peace, health and prosperity throughout the world".

SAFETY REPORTS SERIES No. 129

GUIDANCE FOR REVIEWING SITE CHARACTERISTICS IN SAFETY ANALYSIS REPORTS FOR NUCLEAR INSTALLATIONS

INTERNATIONAL ATOMIC ENERGY AGENCY
VIENNA, 2026

© IAEA, 2026

Printed by the IAEA in Austria
February 2026
STI/PUB/2124
https://doi.org/10.61092/iaea.zy7h-igtp

IAEA Library Cataloguing in Publication Data

Names: International Atomic Energy Agency.
Title: Guidance for reviewing site characteristics in safety analysis reports for nuclear
    installations / International Atomic Energy Agency.
Description: Vienna : International Atomic Energy Agency, 2026. | Series: Safety
    reports series, ISSN 1020-6450 ; no. 129 | Includes bibliographical references.
Identifiers: IAEAL 25-01802 | ISBN 978-92-0-127125-9 (paperback : alk. paper) |
    ISBN 978-92-0-127225-6 (pdf) | ISBN 978-92-0-127325-3 (epub)
Subjects: LCSH: Nuclear facilities — Safety measures. | Nuclear facilities — Safety
    regulations. | Nuclear facilities — Design and construction.
Classification: UDC 621.039.58 | STI/PUB/2124

FOREWORD

The IAEA is dedicated to enhancing the safety and security of nuclear installations worldwide. For countries embarking on new nuclear energy programmes, comprehensive safety evaluations and robust regulatory frameworks are essential.

This publication is part of the IAEA's extrabudgetary project for capacity building in site safety review and assessment in embarking countries.

It is aligned with IAEA Safety Standards Series No. SSG-61, Format and Content of the Safety Analysis Report for Nuclear Power Plants, and is intended to support regulatory bodies in their review of safety analysis reports submitted in accordance with SSG-61 and other IAEA Safety Standards Series publications. The primary features of this publication are comprehensive methods and considerations for the review of the site characteristics chapter of the safety analysis report for nuclear installations, including conventional large nuclear power plants, small modular reactors, research reactors and other types of installation, as defined in the IAEA Nuclear Safety and Security Glossary.

This publication will be an essential tool for regulatory bodies involved in the site safety review and site evaluation of nuclear installations. It will also be useful for organizations preparing the site characteristics chapter in the safety analysis report for regulatory review.

The IAEA is grateful to the European Commission, the Government of the United States of America and the Korea Institute of Nuclear Safety for their financial support of this publication. The IAEA also thanks all who contributed to the drafting and review of this publication, in particular A. Gürpinar (Türkiye), as well as the Member State experts who provided valuable feedback on the draft version through a consultancy meeting, hands-on practices, workshops and a technical meeting. The IAEA officer responsible for this publication was H. Lee of the Division of Nuclear Installation Safety.

CONTENTS

1. INTRODUCTION

1.1. BACKGROUND

The IAEA has been implementing extrabudgetary projects for capacity building in site safety review and assessment in embarking countries, aiming to enhance the capacity of the managers and technical personnel within regulatory bodies who are responsible for reviewing and assessing chapter 2 ('Site characteristics') of the safety analysis report (SAR) for nuclear installations since 2020. These projects have specifically targeted Member States embarking on new nuclear programmes, and this publication is part of these capacity building projects.

The IAEA has developed a self-assessment module for site survey, selection and evaluation, accompanied by a framework for conducting training workshops. Additionally, this framework includes an SAR review practice module focused on the chapter concerning site characteristics, comprising SAR review guidance, a sample SAR, a sample safety evaluation report (SER)[1], and templates and procedures for practical application. Seven Member States were selected to review these products on a pilot basis through hands-on training courses conducted at national and interregional workshops.

Originally developed as a training document for the SAR review practice module, the contents of this publication primarily aim to assist nuclear regulatory bodies and their technical support organizations in evaluating the site characteristics chapter within the SAR for nuclear installations. The contents have undergone significant expansion, and the format has been restructured into this IAEA Safety Report.

1.2. OBJECTIVE

The objective of this publication is to provide practical guidance and information for reviewing the site characteristics chapter of the SAR for nuclear installations. The process includes examining site characteristics that could impact the exposure of both people and non-human biota, as well as initiating the development of arrangements for emergency preparedness and response.

---

[1] In some States, 'SER' may stand for either 'site evaluation report' or 'safety evaluation report'. However, in this publication, 'SER' refers exclusively to the 'safety evaluation report', which contains the review results prepared by the regulatory body after reviewing the 'safety analysis report'.

This review process aligns with the contents and format outlined in IAEA Safety Standards Series No. SSG-61, Format and Content of the Safety Analysis Report for Nuclear Power Plants [1]. Guidance and recommendations provided here in relation to identified good practices represent expert opinion but are not made on the basis of a consensus of all Member States.

1.3. SCOPE

The scope of this publication encompasses the process and practical guidance for the review of the site characteristics chapter of the SAR for nuclear installations, including small modular reactors (SMRs), in accordance with SSG-61 [1]; IAEA Safety Standards Series No. SSR-1, Site Evaluation for Nuclear Installations [2]; and other IAEA Safety Standards that provide more detailed technical recommendations. Therefore, the technical staff of regulatory bodies can use this publication as practical guidance for the site safety review for nuclear installations or as a reference for establishing their own review guidelines. For this reason, the main text was written with the assumption that national regulations would adopt SSG-61 [1] for the 'format and content of the SAR for nuclear installations', as well as other IAEA Safety Standards applicable to the site characteristics chapter in the SAR [2–14].

The environmental impact assessment (EIA) report is a separate document and not part of the SAR. However, coordinating the review of both the SAR and the EIA report can be challenging due to their interfaces, and this document addresses these challenges to assist newly engaged States. Consequently, this publication also provides general guidance on reviewing site characteristics related to the EIA.

This review process is typically conducted in stages, starting with a preliminary document review to assess the completeness, adequacy and quality of the SAR, and involves evaluating adherence to a standard format and ensuring that it meets expected quality levels. Therefore, the IAEA has made efforts to provide the Member States with a systematic safety review framework as well as safety related considerations for reviewing the site characterization and evaluation of nuclear installations in the SAR submitted by the applicant.

1.4. STRUCTURE

The publication is structured in two main parts: general aspects, provided in Sections 2 and 3, which are applicable to all subsections of Section 4 of this publication, and section-by-section practical guidance, given in Section 4 of this publication. Both parts provide clarification and follow the format recommended by SSG-61 [1], with reference to general and specific requirements and recommendations from relevant IAEA Safety Standards [3, 4, 7].

The general aspects outlined in Sections 2 and 3 of this publication clarify the expectations of the regulatory body with respect to supplementary information that complements the IAEA Safety Standards, including quantitative regulatory values (e.g. acceptance criteria, dose limits, intervention levels, screening values).

The practical guidance presented in Section 4 of this publication provides detailed instructions and explanations for reviewing each specific topic of the site characteristics chapter in the SAR. Each individual topic is presented in a distinct order: areas of review, acceptance criteria, review procedure, evaluation findings, implementation, detailed review of technical content, reviews for SMRs, and references.

2. GENERAL ASPECTS OF REVIEW OF SITE CHARACTERISTICS IN SAFETY ANALYSIS REPORTS

2.1. ASSUMPTIONS FOR SAFETY OBJECTIVES AND ACCEPTANCE CRITERIA

For the purposes of this Safety Report, it is assumed that the State has safety regulations based on IAEA Safety Standards, as well as safety targets and acceptance criteria.

Normally, the IAEA Safety Standards do not provide numerical acceptance criteria as such criteria depend on the regulatory approach of the State and are closely linked with the general safety targets required by the regulatory body. Acceptance criteria include the following:

(a) Safety objectives, including dose limits to the public, workers and environment in both the short and long term, during normal operation conditions — with reference to national and international standards.
(b) Annual probability of exceedance associated with an external hazard for the design basis. Other levels lower than the design basis may also be required by the regulatory body (e.g. SL-1 and SL-2 earthquakes).
(c) Definition of a beyond design basis external event. The definition may vary depending on the type of hazard.
(d) Screening probabilities for external events.

(e)   Prescribed values for various zones around the nuclear installation (e.g. exclusion area boundary, low population zone).

(f)   Other prescribed zones around the nuclear installation as needed (e.g. no-fly zones).

2.2. LICENSING PROCESS

IAEA Safety Standards Series publications contain recommendations on the licensing process for a nuclear installation (e.g. IAEA Safety Standards Series No. SSG-12, Licensing Process for Nuclear Installations [4]). However, national regulations and practices may vary, and therefore the timing of the review of chapter 2 ('Site characteristics') of the SAR may show differences between States. In a newcomer country, there may be a site and construction licence, but there may also be intermediate licensing or permitting steps. Early in the process of licensing a nuclear installation, an EIA report needs to be prepared.

The IAEA Safety Standards Series publications that can be used as references for reviewing the EIA report are SSR-1 [2] and IAEA Safety Standards Series No. GSG-10, Prospective Radiological Environmental Impact Assessment for Facilities and Activities [8]. Protection of the environment from radiation hazards is part of the fundamental safety objective in IAEA Safety Standards Series No. SF-1, Fundamental Safety Principles [5].

In some States, the review of the complete EIA report might be done by a different authority from the regulatory body. Therefore, depending on the regulatory regime of the State, there may be duplicate reviews of the radiological part of the EIA report and this needs to be taken into account. Appropriate coordination between the different regulatory authorities reviewing the EIA report would be advisable.

For an operating nuclear installation, a safety review is performed periodically (in general every 10 years). If the review of the site related aspects of nuclear safety is being performed as part of the periodic safety review of the nuclear installation, the process described in IAEA Safety Standards Series No. SSG-25, Periodic Safety Review for Nuclear Power Plants [6], needs to be followed.

2.2.1. Review process

The review process needs to include the following steps in compliance with IAEA Safety Standards Series No. GSG-13, Functions and Processes of the Regulatory Body for Safety [7]. A general review and assessment process is presented in para. 3.162 of GSG-13 [7], as follows:

"The review and assessment process should include the following steps:

(a)   Definition of the scope of the review and assessment process;
(b)   Specification of the purpose of and technical bases for the review and assessment process (these could be considered acceptance criteria for the review and assessment);
(c)   Identification of additional information, if necessary, for the review and assessment;
(d)   Performance of a step-by-step review and assessment procedure to determine whether the applicable safety objectives and regulatory requirements have been met for each aspect or topic;
(e)   Decisions [footnote omitted] on the acceptability of the authorized party's safety arguments or the need for further submissions;
(f)   Reporting and documentation."

2.2.2.   Scope of the review

The review will include the site characteristics chapter of the SAR, as well as any technical annexes to the site characteristics chapter of the SAR, such as topical reports. It also includes the section of the EIA report that deals with site characteristics relevant to the radiological impact of nuclear installations.

2.2.2.1.   *Safety analysis report and technical annexes*

SSG-61 [1] includes a list of IAEA Safety Standards Series publications that are applicable to the site characteristics chapter of the SAR and that need to be taken into account in its preparation. Depending on the practice of the State, the information needed for compliance with national regulations may be presented in the SAR. The more detailed analyses necessary for compliance, such as topical reports, may be presented in annexes. The review process needs to cover the SAR as well as all of the associated annexes.

2.2.2.2.   *Site characteristics in the environmental impact assessment report*

In addition to the IAEA Safety Standards Series publications listed in SSG-61 [1], the publications GSG-10 [8] and IAEA Safety Standards Series No. NS-G-3.2, Dispersion of Radioactive Material in Air and Water and Consideration of Population Distribution in Site Evaluation for Nuclear Power Plants [9], are relevant to the EIA report.

2.2.3. Output of the review

2.2.3.1. Safety evaluation report

The SER for the site characteristics chapter of the SAR needs to address the requirements and recommendations of all IAEA Safety Standards Series publications listed in SSG-61 [1], as well as of GSG-10 [8] and NS-G-3.2 [9]. The SER is generally a concise document summarizing the compliance of the SAR with the applicable IAEA safety standards, detailing any deviations and non-conformances.

2.2.3.2. Evaluation of the site characteristics in the environmental impact assessment report

The way in which the results of the review of site characteristics that contribute to the radiological impact on the region and the way in which the assessment results in the EIA report are documented depends on the regulatory process in each State.

2.3. HIGH LEVEL REVIEW: CHECKING FOR COMPLETENESS

The review of the SAR is performed in two steps. The first involves the general quality of the SAR and its completeness. Only if this high level review results in a positive outcome is a detailed review then performed. The format and content of the SAR outlined in SSG-61 [1] are key for assessing completeness. If a major section of the SAR is absent without an acceptable reason provided by the applicant, then a detailed review of the SAR cannot be made. Moreover, if the information is present but the format used does not comply with SSG-61 [1] (with no acceptable reason), or if the information provided is poorly communicated and it is not possible to understand the contents, a detailed independent review also cannot be made. In these cases, the SAR is rejected without going through a detailed review process.

It is expected that the subchapters for site characteristics in the SAR (including the technical attachments) will contain a level of detail comparable to that recommended in the applicable IAEA Safety Standards Series publications, including SSG-6 [1], and will be structured accordingly. It is anticipated that subchapters will include information on the following subjects:

(a) Site description (including site protection structures and ultimate heat sink (UHS));

(b) Description of the regional characteristics relevant to the site hazards;

(c) Development of the database for and data processing of site characteristics;

(d) Hazard evaluation (for hazards applicable to the site(s));

(e) Hazard evaluation results needed by designers to develop the design bases;

(f) Descriptions of other site characteristics relevant for design basis (e.g. geotechnical);

(g) Descriptions of technical attachments such as topical reports;

(h) Descriptions of how the regulatory pyramid has been applied (e.g. national regulations, IAEA safety standards, third country (such as vendor country) regulations, industry standards);

(i) Technical references (only those that can be provided to the reviewer are to be cited).

The main text of the SAR needs to contain sufficient detail for reviewers to evaluate it. However, detailed calculations and data sets are expected to be presented in technical reports or appendices to the SAR.

In this regard, para. 2.8 of SSG-61 [1] states:

"The preliminary safety analysis report should contain sufficiently detailed information, specifications and supporting calculations to assess and demonstrate that the plant can be constructed, commissioned, operated and decommissioned in a manner that is acceptably safe throughout its lifetime. The preliminary safety analysis report should demonstrate that the requirements specified in the initial safety analysis report are met. The safety features incorporated into the design should be described, with due regard to any site specific aspects. [footnote omitted]"

2.4. REVIEW OF TOPICS RELATED TO EXCLUSIONARY CRITERIA

The SAR reflects the site characterization work shown in fig. 1 of IAEA Safety Standards Series No. SSG-35, Site Survey and Site Selection for Nuclear Installations [10], including the verification of site acceptability. For this reason, the review prioritizes the site acceptability issues (based on exclusionary criteria). It is important that such issues be addressed early in the site characterization process, and therefore they need not be deferred to later stages of the review (see Requirement 4 of SSR-1 [2]). Site acceptability issues may be addressed by field investigations, and their review warrants site visits. It is also possible that evidence related to these issues may be time sensitive. One such example is the inspection of trenches to address the issue of fault capability, which requires timely visits and reviews.

2.5. SCOPE OF THE SAFETY ANALYSIS REPORT

The scope of the SAR needs to be well defined from the start. If more than one installation will be constructed simultaneously (or if one will be constructed after another over a short period), the total radiological impact needs to be considered. Furthermore, on-site and off-site emergency preparedness and response arrangements need to consider the possibility of more than one installation being in accident and/or severe accident conditions.

If existing nuclear installations are collocated at the same site, the interaction between these and the new installation needs to be checked (see Requirement 9 of SSR-1 [2]). Older installations may be associated with higher risks, and their failure may constitute an additional hazard for the new installation.

The multiple installation aspect of the site needs to be taken into consideration, possibly using lower tier IAEA publications (i.e. lower than Safety Guides).

For some external hazards, there may be safety related site protection features in place (e.g. sea walls, dykes). Although these features are outside the boundary of the nuclear installation, they would be within the scope of the SAR. Similarly, infrastructure needed for on-site emergency response, as well as cooling related structures, which may be outside the nuclear installation, are also within the scope of the SAR.

2.6. EVOLUTION OF THE SAFETY ANALYSIS REPORT

The SAR is a live document that evolves as site characteristics, design parameters and methods change over time. This changes may be the result of an external review (e.g. Integrated Regulatory Review Service, Operational Safety Review Team service, Site and External Events Design Review Service, Technical Safety Review Service), an external event affecting the nuclear installation, a large external event occurring at a different location that provides new data for hazard evaluation, or new scientific findings that may impact hazard evaluation. It is important to reflect such changes into the revision of the SAR for future reviews (e.g. the periodic safety review).

2.7. SCOPE OF THE ENVIRONMENTAL IMPACT ASSESSMENT

Depending on the practice of a State, establishing the scope of the EIA may be the first step in the licensing process and will be based on the type of nuclear installation for which the EIA report is being prepared. Defining the licensing sequence for a nuclear installation can also be challenging. For example, if an EIA report is needed before a site licence is granted by the regulatory body, there may be uncertainty regarding the type and number of nuclear installations to be constructed. This uncertainty can lead to a conservative enveloping approach. NS-G-3.2 [9] provides recommendations that can serve as a basis for such an approach.

The Section of the EIA report addressing site characteristics that contribute to the radiological impact on the region needs to include the following basic information:

(a)   Environmental background, including population distribution;
(b)   Analysis of dispersion of radionuclides in the atmosphere;
(c)   Analysis of transport of radionuclides in surface water;
(d)   Analysis of transport of radionuclides in groundwater;
(e)   Assessment of the overall radiological impact;
(f)   Monitoring of radioactivity in the environment;
(g)   Consideration of the feasibility of effective emergency response actions.

Depending on the nuclear installation under consideration, a graded approach might be applicable. IAEA Safety Standards Series publications outline general requirements and recommendations on the graded approach, including Requirement 3 of SSR-1 and specific paragraphs such as paras 1.8, 1.13 and 9.7 of IAEA Safety Standards Series No. SSG-9 (Rev. 1), Seismic Hazards in Site Evaluation for Nuclear Installations [12]; paras 1.12, 10.1 and 10.2 of IAEA Safety Standards Series No. SSG-18, Meteorological and Hydrological Hazards in Site Evaluation for Nuclear Installations [13]; and paras 1.18 and 7.2 of IAEA Safety Standards Series No. SSG-21, Volcanic Hazards in Site Evaluation for Nuclear Installations [11].

The scope of the EIA report may also include items not listed above. For example, the scope may include external hazards that are not listed above but are part of nuclear safety and fall within the purview of the regulatory body. Therefore, close coordination may be necessary between regulatory authorities for the review of the EIA report.

2.8. SITE–DESIGN INTERFACE

In general, a preliminary safety analysis report (PSAR) can be written when the design of the installation is well developed. Therefore, the site–design interface is taken into consideration in the safety analysis by linking the site characteristics chapter of the SAR with other chapters of the PSAR. However, there may be cases where a specific design might not be available when the applicant wishes to apply for a licence for the site and instead provides the basis from which the design parameters can be derived. In this case, an enveloping approach may be used, where the maximum parameters of the planned nuclear installation (i.e. number and installed power of units, footprint, source terms) are assumed. All site characteristics that subsequently change due to newly discovered information, new methodologies and the site–design interface during the construction and/or operation of the nuclear installation need to be incorporated into relevant documents, such as the final SAR, the radiological EIA report and/or the periodic safety review report.

When importing a nuclear installation with a unified design and predefined design bases for external hazards, it is important to verify that all parameters related to these hazards are addressed in the standard design. Additionally, it is crucial to verify that there is a sufficient margin between the calculated hazard values and the design basis parameters.

2.9. CROSS-CUTTING SUBJECTS AND THEIR REVIEW PROCESS

Although the review process considers each topic in accordance with the format of the SAR, often there are related subjects that need to be reviewed jointly. For example, seismic events may be associated with a significant number of concurrent hazards, such as vibratory ground motion, fault displacement, liquefaction, slope instability, tsunamis and floods due to dam breaks. Parameters that are used in the evaluation of these hazards need to be consistent with one another, and this is a key aspect for the review of the SAR. Furthermore, the reviewers of these hazards may be different, in which case interdisciplinary coordination between multiple experts is needed.

2.10. REFERENCING TOPICS WITH OTHER PARTS OF THE SAFETY ANALYSIS REPORT

The site characteristics chapter of the SAR has a significant number of links with other chapters. In fact, all design parameters for external events are based on the hazard analyses in the site characteristics chapter. Moreover, population characteristics and radionuclide dispersion in atmospheric and aquatic media have strong links with dose calculations and emergency plans. These links need to be indicated in the site characteristics chapter so that the values derived in that chapter can be easily traced throughout the SAR. A table that provides the results of the hazard evaluation and a reference to the relevant chapter and section of the SAR where these results are used is useful and needs to be prepared by the applicant.

At a minimum, the following chapters of the SAR need to be referenced in the site characteristics chapter (chapter 2):

(a) Chapter 1: Introduction and general considerations;
(b) Chapter 3: Safety objectives and design rules of structures, systems and components;
(c) Chapter 9: Auxiliary systems and civil structures;
(d) Chapter 12: Radiation protection;
(e) Chapter 15: Safety analysis;
(f) Chapter 17: Management for safety;
(g) Chapter 19: Emergency preparedness and response;
(h) Chapter 20: Environmental aspects;
(i) External events probabilistic safety assessment.

In addition, when external loads are considered for different nuclear installation systems in the relevant chapters of the SAR (i.e. chapters 4–10 and 11), it is important that appropriate links also be established with these chapters.

2.11. FORMAT DIFFERENCES BETWEEN SSG-61 AND SITE EVALUATION STANDARDS

The format of this Safety Report is consistent with SSG-61 [1], which provides the format of the SAR. In relation to the site characteristics chapter of the SAR, this consistency may result in variations in the titles, sequence and contents compared with those used in the IAEA Safety Standards Series publications for site evaluation. For example, in the Safety Standards Series publications related to site evaluation, topics related to external hazards (i.e. impact of the environment on the nuclear installation) are clearly separated from those related to population and dispersion (i.e. impact of the nuclear installation on people and environment). This is not the case for the sections of SSG-61 [1]. In this Safety Report, this issue is addressed by discussing all external hazards — including their interrelationship — in Section 3, then discussing the review of the database and the analysis of each hazard in the subsequent sections.

Although there is generally good consistency between SSR-1 [2] and SSG-61 [1], some exceptions may exist. For example, some of the Safety Guides related to site evaluation were produced at a different time from SSR-1 [2] and therefore include some differences from the SSR-1 [2] requirements. Requirement 11 of SSR-1 [2], for instance, addresses the safety of the UHS: however, this topic is not addressed in any of the Safety Guides. In addition, actions related to management systems are clearly required in SSR-1 [2] and recommended in the Safety Guides, but there is no explicit mention of such actions in SSG-61 [1].

2.12. LEVEL OF DETAIL OF THE REVIEW

The level of detail in the review of the SAR depends on several major factors. The first is the regulatory framework of the State: prescriptive versus performance based, or a combination of the two. Reviews of studies, complying with performance based requirements, may need a more detailed review, as the author of the work has the freedom to select and justify the methods, assumptions and analytical approaches used to demonstrate compliance with the regulatory requirements. This will put the responsibility on the reviewer to check the validity of the methods and their domains of application.

Other factors that may affect the level of detail of the review include:

(a) Impact of the reviewed item on nuclear safety;
(b) The possibility of a one-time review only (e.g. the review of an excavation);
(c) Complexity and novelty of the subject matter;
(d) Experience of the performer;
(e) Robustness of the quality assurance (QA) programme through which the SAR has been prepared (including a peer review by the applicant).

Accordingly, the review may include spot checks of calculations, repeats of calculations or calculations using different software. As part of the review of the QA programme, the validation and verification of the software need to be checked to ensure that it is appropriate for its intended application.

The regulatory body may wish to develop its own procedures for the review of field and laboratory investigations of the applicant.

2.13. REGULATORY PYRAMID

The SAR contains a description of the applicable regulations, norms and standards with the applied hierarchy. This means that the regulation at the top of the hierarchy is preferably used unless there are gaps in this regulation. In this case, the regulation (or standard) next in line can be used, and so on. For each subject, the regulation or standard used may change, and such changes are to be clearly indicated in the chapter or subchapter. The general hierarchy would be (i) national regulations, (ii) IAEA safety standards, (iii) regulations of the vendor country and (iv) internationally recognized industry standards. However, this hierarchy may be changed with the approval of the regulatory body.

2.14. OUTPUT OF THE REGULATORY REVIEW

The output of the regulatory review of the SAR will be the SER, which documents the findings of the regulatory body in a succinct report.

The format of the SER is generally similar to the SAR, as recommended in SSG-61 [1]. Within each topical area, a sectional format may be adopted by the regulatory body, comprising:

(a) Basis of the review (i.e. the applicable IAEA safety standards);
(b) Compliance with requirements (e.g. SSR-1 [2]);
(c) Compliance with recommendations (e.g. SSG-21 [11], SSG-18 [13]);
(d) Record of non-compliance and non-conformances;
(e) Requirements for additional clarification or additional work.

The format of the EIA review report needs to be consistent with the format of the EIA report, which is generally provided by the relevant regulatory authority (e.g. the environment ministry). The applicable IAEA safety standards for this review are mentioned in SSR-1 [2] and NS-G-3.2 [9], which provide requirements and recommendations, respectively.

3. COMPLIANCE WITH IAEA SAFETY STANDARDS SERIES No. SSR-1 IN THE REVIEW OF SITE CHARACTERISTICS

3.1. SAFETY REQUIREMENTS AND RECOMMENDATIONS

Sections 4.1–4.11 of this publication provide practical guidance in relation to specific requirements established in SSR-1 [2]. However, there are more general and cross-cutting SSR-1 [2] requirements with which the applicant needs to comply. In this Section of the Safety Report, practical guidance is provided for the SAR review in relation to these general requirements.

Requirement 2 of SSR-1 [2] states that **"Site evaluation shall be conducted in a comprehensive, systematic, planned and documented**

manner in accordance with a management system." Associated requirements for the establishment of an integrated management system, for QA arrangements and for clear acceptance criteria for site evaluation activities, as well as for the documentation of studies and independent reviews, are outlined in paras 3.1–3.5 of SSR-1 [2]. These requirements cover the leadership, planning and control necessary to ensure safety, regulatory compliance and proper site characterization, including through the evaluation of external hazards and radiological impacts.

Site selection and evaluation are some of the earliest activities to be performed by the applicant in a nuclear installation project. For this reason, it is important that an effective management system be established in a timely manner so that site related activities that have nuclear safety implications are conducted in compliance with Requirement 2 of SSR-1 [2].

High level requirements are provided in SSR-1 [2]. In addition to these requirements, each of the more recently published Safety Guides on site evaluation provide recommendations on the establishment of a management system. These recommendations are generally similar to one another and differ only in aspects that are particular to the specific approach related to the hazard assessment in question.

SSG-9 (Rev. 1) [12] contains recommendations, in support of SSR-1 [2], of the common elements that are expected to be included in a management system for site evaluation activities. Paragraph 10.2 of SSG-9 (Rev. 1) [12] states:

"A project work plan should be established that, at a minimum, addresses the following topics:

(a)   The objectives and scope of the project;
(b)   Applicable regulations and standards;
(c)   Organization of the roles and responsibilities for management of the project;
(d)   Work breakdown, processes and tasks, and schedule and milestones;
(e)   Interfaces among the different tasks (e.g. field tasks, laboratory tests, analysis) and disciplines (e.g. earth sciences, engineering) involved, with all necessary inputs and outputs;
(f)   Project deliverables and reporting."

More generally, paras 10.1–10.19 of SSG-9 (Rev. 1) [12] provide recommendations on the application of the management system in relation to project management, engineering uses and output specification, and independent peer review.

Paragraph 10.5 of SSG-9 (Rev. 1) [12] outlines requirements for clearly identifying methodologies, describing expert interactions and specifying the roles and responsibilities of experts in the project work plan.

Paragraph 10.8 of SSG-9 (Rev. 1) [12] emphasizes that the hazard assessment documentation needs to provide a clear and comprehensive description of all project elements to ensure transparency and traceability for stakeholders such as peer reviewers, regulatory bodies and contractors. This includes detailing the roles of participants, providing analysis documentation with raw and processed data, describing the software and associated input/output files, referencing supporting materials, addressing uncertainties and expert opinions, and presenting results from intermediate calculations and sensitivity studies.

Paragraph 10.12 of SSG-9 (Rev. 1) [12] highlights that seismic hazard assessments for nuclear installations are typically carried out to support seismic design and probabilistic safety evaluations. From the outset, the work plan for such assessments identifies the engineering purposes and objectives, specifying the required outputs, including all results that are essential for achieving these purposes and objectives.

Paragraph 10.15 of SSG-9 (Rev. 1) [12] emphasizes the importance of incorporating independent peer review into the project work plan for seismic hazard assessments. This review process ensures that the analysis follows an appropriate methodology, adequately addresses and evaluates both epistemic and aleatory uncertainties, and produces documentation that is complete and traceable.

The reviewer needs to verify that all of the necessary management system documents are in place and are being implemented as part of the site evaluation process.

Requirement 3 of SSR-1 [2] states:

"The scope of the site evaluation shall encompass factors relating to the site and factors relating to the interaction between the site and the installation, for all operational states and accident conditions, including accidents that could warrant emergency response actions."

Associated requirements for the interaction between the site and the installation outline that site evaluation includes external hazards, monitoring activities and site specific parameters, using a graded approach based on radiation risks. The level of detail varies with the potential hazards and risks of the installation and its site, with the scope tailored to support the safety demonstration, as outlined in paras 4.1–4.4 of SSR-1 [2].

Requirement 3 in SSR-1 [2] has the objective of ensuring completeness. Checking the completeness of the SAR is the initial action of a reviewer before deciding whether the SAR can be reviewed. The information provided in Requirement 3 of SSR-1 [2] on the scope and content of site evaluation is considered a prerequisite to completeness and needs to be verified by the reviewer at the beginning of the review process.

Compliance with the format recommended in SSG-61 [1] will facilitate completeness.

Requirement 4 and the associated paras 4.6–4.11 of SSR-1 [2] state:

"The suitability of the site shall be assessed at an early stage of the site evaluation and shall be confirmed for the lifetime of the planned nuclear installation.

4.6. In the assessment of the suitability of a site for a nuclear installation, the following aspects shall be addressed at an early stage of the site evaluation:

(a) The effects of natural and human induced external events occurring in the region that might affect the site;

(b) The characteristics of the site and its environment that could influence the transfer of radioactive material released from the nuclear installation to people and to the environment;

(c) The population density, population distribution and other characteristics of the external zone, in so far as these could affect the feasibility of planning effective emergency response actions [15], and the need to evaluate the risk to individuals and to the population.

4.7. The site shall be deemed unsuitable for a nuclear installation if one or more of the three aspects listed in para. 4.6 indicates that the site is unacceptable and the deficiencies cannot be compensated for by means of a combination of measures for site protection, design features of the nuclear installation and administrative procedures.

4.8. Site suitability shall be assessed on the basis of relevant current data and methodologies. If relevant, conservative criteria shall be developed in relation to site specific accident scenarios, and the consistency of such criteria with the overall site suitability shall be demonstrated.

4.9. A decision regarding the suitability of the site shall be based on the characteristics of the nuclear installation, including planned operations at the site, the amount and nature of potential radioactive releases and their impact on people and the environment.

4.10. For nuclear power plants, the total nuclear capacity to be installed at the site shall be determined at the first stages of the siting process. If it is later determined or anticipated that the installed nuclear capacity (or, for other nuclear installations, the inventory of nuclear material) or its impact has increased to a level significantly greater than that previously determined to be acceptable, the site shall be re-evaluated considering the higher capacity, inventory or impact.

4.11. In the overall evaluation of site suitability, site specific attributes, such as cooling water availability or extreme environmental conditions, and their potential role in affecting the safe and continuous operation of the nuclear installation, shall also be addressed."

In accordance with SSR-1 [2] and SSG-35 [10], at the beginning of the site evaluation process, it is necessary to revisit site suitability and confirm that there are no exclusionary attributes that make the site unsuitable for the safe operation of the nuclear installation.

The site suitability review is a high priority task and needs to be addressed early in the site evaluation process. Furthermore, by the time the SAR is prepared, a clearer indication of the feasibility of engineering solutions to site related difficulties would be expected, and these solutions need to be carefully reviewed. Sufficiently conservative margins are expected to be present in the engineering measures when dealing with hazards or phenomena that may impact the site suitability.

Although site suitability may seem to be an issue for new sites only, it may also be an issue when a new installation is planned to be collocated with an older one, or when an existing facility is being expanded. Some exclusionary attributes may be very local (e.g. geotechnical issues such as liquefaction or karst collapse) and may concern only the planned installation. Even when an SAR is updated for an operating nuclear installation (e.g. in the framework of a periodic safety review), site suitability may emerge as an issue due to a newly discovered feature or a new development related to the site characteristics.

Requirement 5 of SSR-1 [2] states that "**The site and the region shall be investigated with regard to the characteristics that could affect the safety of the nuclear installation and the potential radiological impact of the nuclear installation on people and the environment.**" This requirement is addressed in Sections 4.1–4.11 of this Safety Report.

Requirement 6 of SSR-1 [2] states that "**Potential external hazards associated with natural phenomena, human induced events and human activities that could affect the region shall be identified through a screening process.**"

Requirement 7 of SSR-1 [2] states that "**The impact of natural and human induced external hazards on the safety of the nuclear installation shall be evaluated over the lifetime of the nuclear installation.**"

Requirement 8 of SSR-1 [2] states that "**If the projected design of the nuclear installation is not able to safely withstand the impact of natural and human induced external hazards, the need for site protection measures shall be evaluated.**" Associated requirements are established in paras 4.29–4.31 of SSR-1 [2].

Site protection measures are discussed Section 4 of this Safety Report. This information needs to be explicitly presented, together with the associated design basis and, if applicable, the beyond design basis. Some site protection features (e.g. dykes) are outside the site area and not necessarily under the supervision of the nuclear installation's operating organization. An account of how these features are to be kept up and maintained is needed.

A topic not explicitly addressed in Requirement 8 of SSR-1 [2] is the administrative measures that may be implemented to deal with human induced hazards. Procedures need to be in place to outline how these administrative measures can be implemented. If these measures need coordination with authorities outside the nuclear installation (e.g. enforcement of a no-fly zone), then evidence of this coordination, with periodic updates, needs to be demonstrated.

Requirement 9 of SSR-1 [2] states that "**The site evaluation shall consider the potential for natural and human induced external hazards to affect multiple nuclear installations on the same site as well as on adjacent sites.**" Associated requirements are established in paras 4.32 and 4.33 of SSR-1 [2].

Many existing nuclear installation sites host more than one unit, and new units are frequently planned to be collocated with older units. One important lesson from the accident at the Fukushima Daiichi nuclear power plant relates to multiple unit sites and the potential for an accident at one unit to affect the safety of a collocated unit. The common cause and effect of large external hazards (e.g. earthquakes, volcanoes, flooding) needs to be considered in the safety analysis of the collocated units. Safety systems that may be shared between units need particular attention. Another issue arises when the same site hosts multiple units of different ages, which may lead to the possibility of a single unit failing and becoming a source of 'human-induced hazard' to the other units. The reviewer needs to check that all such potential interactions have been considered with adequate margins.

Requirement 10 of SSR-1 [2] states that "**The external hazards and the site characteristics shall be assessed in terms of their potential for changing over time and the potential impact of these changes shall be evaluated.**" This requirement is addressed in Section 4 of this Safety Report. The assessment referred to in this requirement also needs to consider any changes in methods and tools used to evaluate the external hazards.

Requirement 11 of SSR-1 [2] states:

"The evaluation of site specific natural and human induced external hazards for nuclear installations that require an ultimate heat sink shall consider hazards that could affect the availability and reliability of the ultimate heat sink."

Requirement 12 of SSR-1 [2] states:

"In determining the potential radiological impact of the nuclear installation on the region for operational states and accident conditions, including accidents that could warrant emergency response actions, appropriate estimates shall be made of the potential releases of radioactive material, with account taken of the design of the nuclear installation and its safety features."

Requirement 13 of SSR-1 [2] states:

"The feasibility of planning effective emergency response actions on the site and in the external zone shall be evaluated, with account taken of the characteristics of the site and the external zone as well as any external events that could hinder the establishment of complete emergency arrangements prior to operation."

Requirement 14 of SSR-1 [2] states:

"The data necessary to perform an assessment of natural and human induced external hazards and to assess both the impact of the environment on the safety of the nuclear installation and the impact of the nuclear installation on people and the environment shall be collected."

3.2. SITE EVALUATION FOR INSTALLATIONS OTHER THAN LARGE NUCLEAR POWER PLANTS

The IAEA safety standards for site evaluation are generally written to apply to all nuclear installations, as listed in para. 1.7 of SSR-1 [2], with provisions for the application of a graded approach for installations other than large nuclear power plants, such as SMRs, research reactors, fuel cycle facilities and interim storage facilities. Paragraph 4.5 of SSR-1 [2] states:

"For site evaluation for nuclear installations other than nuclear power plants, the following shall be taken into consideration in the application of a graded approach:

(a)   The amount, type and status of the radioactive inventory at the site (e.g. whether the radioactive material on the site is in solid, liquid and/or gaseous form, and whether the radioactive material is being processed in the nuclear installation or is being stored on the site);

(b)   The intrinsic hazards associated with the physical and chemical processes that take place at the nuclear installation;

(c)   For research reactors, the thermal power;

(d)   The distribution and location of radioactive sources in the nuclear installation;

(e)   The configuration and layout of installations designed for experiments, and how these might change in future;

(f)   The need for active systems and/or operator actions for the prevention of accidents and for the mitigation of the consequences of accidents;

(g)   The potential for on-site and off-site consequences in the event of an accident."

Recommendations on applying a graded approach are provided in IAEA Safety Guides, such as SSG-9 (Rev. 1) [12], SSG-18 [13] and SSG-21 [11]. It is important to apply a graded approach with care. For example, the use of a smaller database and simpler methods would need to be balanced by allowing higher annual probabilities of exceedance to be permitted in the hazard analyses.

3.3. ADDITIONAL CONSIDERATIONS FOR SMALL MODULAR REACTORS

If smaller nuclear power plants, such as SMRs, are proven to present a lower risk due to their enhanced safety features and their smaller source term compared with large nuclear power plants, then a graded approach can be used when applying the safety requirements and recommendations, and requirements and recommendations may even possibly be revised to reflect this. In 2024, the IAEA published IAEA-TECDOC-2042, Optimization of Safety Measures for Protection of Nuclear Installations Against External Hazards [16]. In that publication, practical guidance is offered on developing a technology-neutral safety framework for site evaluation that considers site–installation interactions and the innovative safety features of advanced reactors.

To focus on the more concrete elements of the grading process, the most recently published Safety Guide related to site evaluation is used. Section 9 of SSG-9 (Rev. 1) [12] provides recommendations on the application of a graded approach in the evaluation of seismic hazards. These recommendations provide the basis for applying a graded approach to the issues addressed in this Safety Report, particularly para. 9.5 of SSG-9 (Rev. 1) [12] on the screening process and paras 9.8 and 9.9 of SSG-9 (Rev. 1) [12] on the categorization process.

Paragraph 9.5 of SSG-9 (Rev. 1) [12] outlines some relevant factors relating to the characteristics of nuclear installations, as follows:

"(a) The amount, type and status of the radioactive inventory at the site (e.g. whether solid, liquid and/or gaseous; whether the radioactive material is being processed or only stored);

(b) The intrinsic hazard associated with the physical processes (e.g. nuclear chain reactions) and chemical processes (e.g. for fuel processing purposes) that take place at the installation;

(c) The thermal power of the nuclear installation, if applicable;

(d) The configuration of the installation for different kinds of activity;

(e) The distribution of radioactive sources in the installation (e.g. for research reactors, most of the radioactive inventory will be in the reactor core and the fuel storage pool, whereas for fuel processing and storage facilities it might be distributed throughout the installation);

(f) The changing nature of the configuration and layout of installations designed for experiments (such activities have an associated intrinsic unpredictability);

(g) The need for active safety systems and/or operator actions for the prevention of accidents and for mitigation of the consequences of accidents, and the characteristics of engineered safety features for the prevention of accidents and for mitigation of the consequences of accidents (e.g. the containment and containment systems);

(h) The characteristics of the structures of the nuclear installations and the means of confinement of radioactive material;

(i) The characteristics of the processes or of the engineering features that might show a cliff edge effect in the event of an accident;

(j) The characteristics of the site that are relevant to the consequences of the dispersion of radioactive material to the atmosphere and the hydrosphere (e.g. size and demographics of the region);

(k) The potential for on-site and off-site contamination."

As a consequence of the screening process, a categorization of the SMR is undertaken.

Paragraph 9.8 of SSG-9 (Rev. 1) [12] states:

"If the conservative screening process indicates that a seismic hazard assessment of the installation is to be carried out (see para. 9.5), a process for categorizing the installation should be undertaken. This categorization may be performed at the design stage or later. If the categorization has been performed, the assumptions on which it was based should be reviewed and verified. In general, the criteria for categorization should be based on the radiological consequences of a radioactive release from the installation, ranging from very low to potentially severe consequences. As an alternative, the categorization may consider the radiological consequences within the installation itself, within the site of the installation, and for the public and the environment."

Paragraph 9.9 of SSG-9 (Rev. 1) [12] states:

"Three or more categories may be defined on the basis of national practice and criteria, as well as the information described in para. 9.7. As an example, the following categories may be defined:

"(a) The lowest hazard category, which includes those nuclear installations for which national building codes for conventional installations (e.g. essential facilities such as hospitals) or for hazardous facilities (e.g. petrochemical or chemical plants) should be applied as a minimum;

(b) The highest hazard category, which includes installations for which standards and codes for nuclear power plants should be applied;

(c) There is often at least one intermediate category between (a) and (b), corresponding to a hazardous installation for which, at a minimum, codes dedicated to hazardous facilities should be applied."

3.3.1. Grading in relation to non-exclusionary attributes

Paragraph 9.10 of SSG-9 (Rev. 1) [12] outlines recommendations for grading non-exclusionary attributes in the vibratory ground motion hazards assessment. These recommendations can also be applied to all other non-exclusionary attributes. As stated in para. 9.10 of SSG-9 (Rev. 1) [12], the hazard evaluation for installations categorized as suggested in paras 9.8 and 9.9 of SSG-9 (Rev. 1) [12] needs to follow these recommendations:

"(a) For the least hazardous installations, the input ground motion for the design may be taken from national building codes and maps.

(b) For installations in the highest hazard category, methodologies for seismic hazard assessment as described in Sections 3–8 of this Safety Guide (i.e. recommendations applicable to nuclear power plants) should be used.

(c) For installations categorized in the intermediate hazard category, the following approach might be applicable:

(i) If the seismic hazard assessment is typically performed using methods similar to those described in this Safety Guide, a lower input ground motion than that evaluated for (b) may be adopted for designing these installations, in accordance with the safety requirements for the installation.

(ii) If the database and the methods recommended in this Safety Guide are found to be disproportionately complex, time consuming and demanding for the nuclear installation in question, simplified methods for seismic hazard assessment (that are based on a more restricted data set) may be used. In such cases, the input ground motion finally adopted for designing the installation should be commensurate with the reduced database and the simplification of the methods, with account taken of the fact that both factors tend to increase uncertainties."

Recommendations on related monitoring instrumentation installed on the site need to be applied in a manner commensurate with the category of the installation, as defined above.

3.3.2. Grading in relation to exclusionary attributes

Paragraph 9.13 of SSG-9 (Rev. 1) [12] provides recommendations on grading exclusionary attributes in seismic and fault displacement hazard assessments. These recommendations can be expanded to all exclusionary attributes. Regarding the exclusionary characteristics of a site, the same principles applied to large nuclear power plants also need to be considered for SMRs. If credible evidence suggests that there is potential for exclusionary attributes within the site vicinity or site area, a thorough and detailed hazard assessment needs to be performed. In cases where engineering measures can effectively mitigate exclusionary attributes, the site may still meet suitability requirements under clearly defined criteria. In such scenarios, appropriate design bases need to be developed to ensure the nuclear installation's safety through measures related to design, construction and operational practices.

3.3.3. Types of design for small modular reactors

Six types of SMR are included in the IAEA SMR Book of 2024 [17]. These, together with the number of active designs with demonstrated sustained development, are listed as follows:

(a) Water cooled, land based SMRs (14);
(b) Water cooled, marine based SMRs (6);
(c) High temperature, gas cooled SMRs (14);
(d) Fast neutron spectrum SMRs (10);
(e) Molten salt SMRs (11);
(f) Micro-sized SMRs (13).

It is important to note that there is variability in design in terms of:

(a) Power;
(b) Innovative features (as opposed to evolutionary features), including fuel;
(c) Maturity of design;
(d) Utilization (electricity to the grid, heat generation, dedicated power source).

In this Safety Report, it is assumed that the SMR under review has been categorized accurately and that it is either in Category B (i.e. the highest hazard category, to be treated as a large nuclear power plant) or in Category C (i.e. the intermediate hazard category, for which a graded approach may be applied).

In Sections 4.1–4.11 of this Safety Report, suggestions will be made for the possible revision of requirements and recommendations for those SMRs in Category C. The balance between the volume and scope of data collected and the complexity of the approaches used needs to be considered when determining the permissible increase in the annual probability of exceedance for the design basis level of the hazard in question. The ongoing monitoring of these hazards will follow a similar approach. For issues related to the impact of the SMR on people and the environment, the requirements established in SSR-1 [2] will be used for the application of a graded approach. There is no one way of applying a graded approach to requirements and recommendations. Therefore, the reviewer needs to keep in mind the necessary balance in applying this approach.

In this Safety Report, it is assumed that the 'modular' nature of the SMR provides important safety and operational advantages due to the independence of each module and that the operating organization needs to ensure that the likelihood of the failure of multiple modules is extremely low and can be considered impossible. A screening probability applicable to external hazards may be used for this extremely low likelihood. This would be in the order of probabilities used in the application of 'practical elimination' in the safety analysis. Only under this assumption can emergency zone requirements be adjusted to be commensurate with the failure of a single module leading to the consideration of the corresponding source term.

This assumption leads to the conclusion that the probability of failure of the SMR due to an external hazard will be less than that of a hypothetical large nuclear power plant at the same site. However, given the novel deployment scenarios for SMRs, including potential urban area installations, it is crucial to extensively assess new or more prevalent hazards. Consequently, while the site characteristics database might be decreased and less sophisticated methods might be used for hazard analysis in some cases, the resources allocated for site evaluation and hazard analysis still need to be sufficient for a comprehensive risk assessment and to ensure safety.

Site evaluation activities needed for large nuclear power plants could be reviewed to explore the potential for performance based reductions for certain external hazards if an evaluation of the external hazard is irrelevant or unsuitable for the chosen SMR site.

4. REVIEW OF SPECIFIC SITE CHARACTERISTICS

4.1. GEOGRAPHY AND DEMOGRAPHY

4.1.1. Areas of review

The target of the SAR review in this Section of the Safety Report includes the information recommended in paras 3.2.7–3.2.9 of SSG-61 [1]. The purpose of the review in this Section of the Safety Report includes both a review of the potential impact of the surrounding facilities and activities on the safety of the nuclear installation and a review of the potential impacts of the nuclear installation on the people and environment due to planned discharges and accidental releases.

Geographical coordinates need to be provided for each nuclear installation at the site. In particular, the coordinates of the centre of each reactor building need to be identified. Furthermore, the area within the administrative control of the operating organization needs to be indicated on a map of appropriate scale. This information will be used in the identification of the site vicinity, near region and the region for the analysis of external hazards.

Population data are to be collected and presented in a suitable format and scale that enable the data to be correlated with other data sets, such as wind directions and uses of land and water. The data are presented both in tabular and graphical form, generally concentric circles that would match with wind distributions in wind rose diagrams. Both permanent and temporary populations are to be considered.

Requirements 4, 5, 12–14, 26 and 27 of SSR-1 [2] are the most relevant for these areas of the review. Critical groups associated with the nuclear installation are identified, together with their particular dietary habits (see paras 5.11 and 5.12 of NS-G-3.2 [9]).

The census used for population statistics needs to be indicated and necessary extrapolations made for projecting these statistics over the lifetime of the nuclear installation, taking into account the possible impact of the construction of the installation.

Section 4.1 interfaces with Sections 4.2–4.4 and 4.8–4.11 of this Safety Report.

The paragraphs in SSG-61 [1] relevant to these areas of the review state:

"3.2.7. This Section should specify the site location, including both the area under the control of the operating organization and the area surrounding the site in which there is a need for consultation with interested parties on the control of activities that could affect plant operation (e.g. aircraft flights, associated flight exclusion zones). This should include facilities and activities in the surrounding area that could pose a hazard to the plant (e.g. pipelines, roadways, waterways).

"3.2.8. Information on activities with the potential to affect plant operation should include relevant data on the population distribution and density (including, where applicable, transient populations) and on the distribution of public and private facilities (e.g. airports, harbours, rail transport centres, pipelines, roadways, waterways, factories and other industrial sites, schools, hospitals, police services, firefighting services, municipal services) around the site.

"3.2.9. This Section should also cover the public uses of the land and water resources in the surrounding area and should include an assessment of any possible interaction with the plant and the implications for off-site protective actions in an emergency."

4.1.2. Acceptance criteria

IAEA safety standards (in particular those found in SSR-1 [2]) do not require specific numerical acceptance criteria in relation to geography or demography. Therefore, national regulations need to be checked for applicable acceptance criteria. Geographically, these criteria could be related to the distance to international borders or possible restrictions on siting nuclear installations on islands or narrow peninsulas. Demographically, there are States that require specific and prescribed demographic zones, such as exclusion areas and low population zones. Also, there may be restrictions on population density in the site vicinity and/or distance to population centres. This would affect the feasibility of an effective emergency plan.

4.1.3. Review procedure

A typical safety review of an SAR consists of a desktop review of documents submitted by the applicant, site visits to confirm consistency between the SAR and the site condition, and independent calculations for verification.

The review is mainly performed from documents. Maps drawn to appropriate scales need to be prepared by the applicant. In particular, the geographical coordinates of key points, such as the centre point of the reactor building, for the installation (or, if there are multiple installations, for each installation) need to be provided.

The site area for the installation needs to be defined in accordance with SSR-1 [2]. This is the area under the administrative control of the operating organization. If there is more than one collocated nuclear installation, they all need to be indicated on the maps.

A site visit is necessary in order to understand the geographical setting and the demographics of the site vicinity. It is also a good way to understand the setting of the site with respect to the site vicinity infrastructure.

Independent checks of some hazard calculations may also be needed particularly when the results, even after the necessary sensitivity and/or uncertainly analysis, indicate that acceptance criteria are met with only small margins.

4.1.4. Evaluation findings

The reviewer's evaluation of site characteristics, in accordance with the relevant regulatory criteria, needs to be documented in the SER, noting any findings that diverge from these criteria. The first step of the review is a decision on the overall quality and completeness of this Section of the SAR. It then needs to be verified that there are no issues related to site suitability.

The documentation of the review needs to include the evaluation of site characteristics with respect to the relevant regulatory criteria, including any prescribed distances and zones. The reviewer is responsible for independently verifying that the contents of the SAR comply with the acceptance criteria and for ensuring that the conclusions of the evaluation report also meet the acceptance criteria. The procedure followed by the reviewer in the evaluation needs to be briefly described.

The SER needs to include short descriptions of the site area and the site vicinity. If an exclusion area and/or a low population zone is identified, then any important infrastructure in these areas needs to be indicated, and any prominent topographical features need to be highlighted.

It is necessary to verify that the applicant has provided sufficient information and that the review and calculations (if applicable) support the conclusions of the SER.

4.1.5. Implementation

The requirements established in SSR-1 [2] and the recommendations provided in NS-G-3.2 [9] need to be used in the evaluations performed by the applicant, together with all applicable national regulations. If national safety targets and acceptance criteria related to geography and demography are complied with using different approaches from those provided in the national regulations, guidelines and IAEA safety standards, the methods proposed by the applicant need to be duly scrutinized.

4.1.6. Detailed review of technical content

Requirement 26 and associated paras 6.8–6.10 of SSR-1 [2] state:

"The existing and projected population distribution within the region over the lifetime of the nuclear installation shall be determined and the potential impact of radioactive releases on the public, in both operational states and accident conditions, shall be evaluated and periodically updated.

"6.8. Information on the existing and projected population distribution in the region, including resident populations and (to the extent possible) transient populations, shall be collected and kept up to date over the lifetime of the nuclear installation. Special attention shall be paid to vulnerable populations and residential institutions (e.g. schools, hospitals, nursing homes and prisons) when evaluating the potential impact of radioactive releases and considering the feasibility of implementing protective actions.

"6.9. The most recent census data for the region, or information obtained by extrapolation of the most recent data on resident populations and transient populations, shall be used in obtaining the population distribution. In the absence of reliable data, a special study shall be carried out.

"6.10.   The data shall be analysed to obtain the population distribution in terms of the direction and distance from the site. This information shall be used to carry out an evaluation of the potential radiological impact of normal discharges and accidental releases of radioactive material, including reasonable consideration of releases due to severe accidents, with the use of site specific design parameters and models as appropriate."

The database of population statistics is important in evaluating the impact of the nuclear installation on people and the environment as well as in planning for emergencies. The data are generally obtained from official central or local government sources. It is important to correlate these data with the meteorological data obtained from the meteorological tower or sodar (sound detection and ranging) for the emergency plan. This is generally done by considering the wind rose, which represents the meteorological data, and the population information that corresponds to each section of the wind rose.

The region of interest for the collection of population data depends on the external zone defined by the applicable regulations. The allowable dose limits to the population and the results of the accident analysis (based on the source term related to the nuclear installation) will determine the size of the external zone. It is important to consider any future nuclear installations that are planned to be built at the same site in this calculation.

Both permanent and transient populations need thorough consideration. Special facilities such as schools, barracks, hospitals, recreational areas and prisons will pose additional challenges for emergency preparedness and response and need to be addressed separately.

To estimate projections of population changes, it is important to obtain information on the economic development of the region of interest. Therefore, any changes planned by the local or national government to the land or water use need to be obtained for use in the projection models.

It is essential that the population information be updated periodically evaluate whether changes in population metrics over time will affect the hazard evaluations done at the time of the nuclear installation's construction. At the time of the preparation of the PSAR, demographic projections need to be made using models based on credible and substantiated assumptions.

These assumptions mainly consist of the plans for the construction of facilities in the region of interest that would affect population growth and distribution. New factories, military facilities and recreational areas may all affect population growth and distribution. Obviously, the construction of the nuclear installation itself is an important factor and needs to be considered in projection models. It is important to check the validity of these models during the lifetime of the nuclear installation, making corrections when necessary.

Uncertainties may be related to the database itself (depending on the accuracy of the data) and the modelling that is performed based on assumptions, which may also include uncertainties. These uncertainties may be managed by constructing several models (e.g. best estimate, optimistic, pessimistic) and checking the results of all of these models and providing contingencies for pessimistic models.

To review the full impact of population growth and distribution, it is important to consider information gathered on the proximity of industrial, transportation and other facilities; on site characteristics and the potential effects of the nuclear installation in the region; and on site related issues in emergency preparedness and response and accident management (see Sections 4.3, 4.8 and 4.10 of this Safety Report).

4.1.7. Reviews for small modular reactors

The population data to be collected and processed for an SMR depend on the external zone that is considered necessary. Typically, this zone will be significantly smaller than that for a large nuclear power plant; however, this depends on the number of modules planned and whether the extent of the external zone is determined conditional to the failure of a single module.

One of the main considerations of a modular design is the premise that the external zone is determined conditional to the failure of a single module. In this case, the reviewer needs to confirm that all common cause scenarios have been taken into account. These scenarios may involve shared systems such as the reactor pool, the spent fuel pool and the control room, as well as large external hazards with the potential for impacting more than one module (e.g. earthquakes, floods).

SMRs have a wide variety of designs, and therefore the requirements for the external zone need to be considered on a case-by-case basis. Moreover, these requirements depend on whether there is sufficient justification for the assumption of the failure of a single module only. If the application of a graded approach to the requirements for the external zone is justified, then the collection and processing of demographic data may involve a much smaller area than for a large nuclear power plant.

The data collected from the region relating to infrastructure and hazardous facilities need to consider the infrastructure necessary for dealing with sheltering and evacuation in the case of a nuclear or radiological emergency. Such data are needed for the establishment of a viable off-site emergency response plan. The extent of these data, while discussed here in the context of SMRs, is linked to the general discussion on demographics provided in the preceding subsections for nuclear installations in general.

Data on infrastructure such as roads, railroads, ports and airports are also important in determining distances for potential sources of human induced hazards. Aside from these 'mobile' types of source, stationary sources such as arsenals, hazardous substance storage facilities, petrochemical facilities and pipelines need to be considered in the database. For all of these sources of human induced hazards, the screening distances are based on the maximum potential of an accident at the source, as well as the design of the SMR with respect to the load case(s) generated by this accident. Typically, it can be assumed that SMRs might not be as robust as a large nuclear power plant for large human induced hazards. This would mean that the data from a comparatively larger area would need to be collected and processed.

4.1.8. References for the review

The references for Section 4.1 include SSG-61 [1], SSR-1 [2], NS-G-3.2 [9] and SSG-35 [10].

4.2. EVALUATION OF SITE SPECIFIC HAZARDS

4.2.1. Areas of review

The target of the SAR review in Section 4.2 includes the information recommended in paras 3.2.10–3.2.17 of SSG-61 [1].

The purpose of Section 4.2 is to address the overall evaluation of site specific hazards and their interrelationships, combinations, target probabilities and acceptance criteria, as well as the screening of events and considerations for multiple events.

The applicant is expected to start with a complete list of natural and human induced external hazards using IAEA Safety Standards Series publications (e.g. SSR-1 [2], SSG-35 [10], SSG-21 [11], SSG-9 (Rev. 1) [12], SSG-18 [13], and IAEA Safety Standards Series Nos SSG-79, Hazards Associated with Human Induced External Events in Site Evaluation for Nuclear Installations [18]; NS-G-3.6, Geotechnical Aspects of Site Evaluation and Foundations for Nuclear Power Plants [19]; and SSG-68, Design of Nuclear Installations Against External Events Excluding Earthquakes [20]) as a basis. Some self-evident hazards may be screened out (e.g. ice hazards in a tropical climate). For others, the application of a formal screening process is expected. For example, if there are no volcanoes within a 300 km radius of the site, it does not mean that volcanic hazards can be screened out without a formal screening process (i.e. it is not self-evident that volcanic hazard is absent). Volcanoes farther away can produce tephra fall that may need to be considered in the design of the nuclear installation (see para. 4.5 of SSG-21 [11]).

All effects from the source of the hazard are considered in the screening process. For example, lava flow from a volcano may reach a distance of only tens of kilometres, but ash fall from the same volcano may spread over a very large area, reaching distances of hundreds of kilometres.

In general, hazards are screened out using distance and probability. First, 'distance' is used to deterministically show that, even if the source produces a hazard of the greatest potential intensity, it cannot possibly affect the safety of the nuclear installation. In such a case, the hazard can be screened out. If it cannot be screened out, screening by probability is used. The screening level is normally prescribed by the regulatory body. If this is not the case, a level several orders of magnitude less than that corresponding to the design basis (i.e. $\sim 10^{-7}$ per year) may be used. Screening by distance is performed for each effect of the source.

All external hazards that have not been screened out need to be considered in the hazard analysis. While the recurrence of some hazards is considered stationary, the recurrence of others (especially human induced hazards) is clearly non-stationary, and therefore their potential for change needs to be taken into account. Although hydrological and rare meteorological hazards may have been modelled as stationary until recently, global warming needs to be considered in their evolution up to the end of the nuclear installation's lifetime.

For each of these hazards, target probability levels (annual probabilities of exceedance) are prescribed by the regulatory body. While the design bases for external hazards are generally defined probabilistically, beyond design bases for these hazards could be defined using different approaches, including probabilistic considerations [20].

Some hazards may be correlated with one another due to causation, and this needs to be considered in the analysis. Also, several hazards need to be combined with ambient conditions or other hazards to be effectively evaluated. A typical example is coastal flooding, where phenomena such as wave action, storm surges, seiches, tsunamis and tides need to be combined to calculate the design basis coastal flood. Probabilistic considerations are used in combining these hazards and phenomena.

Requirements 15–24 in SSR-1 [2] and paras 3.2.10–3.2.17 of SSG-61 [1] are the most relevant for these areas of the review.

Section 4.2 interfaces with Sections 4.5–4.7 and 4.11 of this Safety Report. The paragraphs in SSG-61 [1] relevant to these areas of the review state:

"3.2.10. This Section should present the results of a detailed evaluation of natural and human induced hazards at the site that should be taken into account in the design of SSCs [structures, systems and components]. The description should include due consideration of the envisaged evolution of these hazards during the expected lifetime of the nuclear power plant. SSR-1 [2] establishes requirements for the evaluation of specific external hazards.

"3.2.11. The screening criteria used for each hazard (including the envelope, probability thresholds and credibility of events) and the expected impact of each hazard in terms of the originating source, the potential propagation mechanisms and the predicted effects at the site should be described in this section.

"3.2.12. Hazards identified as potentially affecting the site can be screened out if they would be incapable of challenging the safety of the plant or if they are considered, with a high degree of confidence, to be extremely unlikely. The arguments in support of the screening process should be justified and described in this Section of the safety analysis report.

"3.2.13. The target probability levels for design against external hazards should be defined, and a comparison with the acceptable limits should be presented. Attention should be paid to the external hazards that could lead to common cause failures of the safety systems and the safety features for design extension conditions.

"3.2.14. The evaluation presented in this Section should also take into account unlikely natural hazards exceeding those considered for design, derived from the hazard evaluation for the site, to ensure adequate margins to avoid cliff edge effects. The reliability of the heat transfer to the ultimate heat sink should be given special attention.

"3.2.15. This Section should confirm that appropriate arrangements are in place to periodically update the evaluations of site specific hazards in accordance with the results of updated methods of evaluation, monitoring data and surveillance activities.

"3.2.16. This Section should also include results from the evaluation of potential combinations of site specific hazards that could affect the safety of the nuclear power plant.

"3.2.17. Where administrative measures are employed to mitigate the adverse effects of hazards (especially for human induced events), information should be presented on their implementation, together with the roles and responsibilities for their enforcement."

4.2.2. Acceptance criteria

Acceptance criteria include the probabilistic screening level and target probability values for the design bases and beyond design bases. For the beyond design bases, other approaches are also used, as mentioned in Section 4.2.1 of this Safety Report.

Both the probabilistic screening level and the annual probability of exceedance for external hazards are part of national regulations. If this is not the case, it is possible to use relevant recommendations provided in IAEA Safety Guides, if available. In any event, these levels and probabilities need to be discussed with the regulatory body before proceeding with the hazard analysis. For screening out external hazards, a value of 10^{-7} per year is generally accepted, unless prescribed differently by the regulatory body.

4.2.3. Review procedure

A typical safety review of an SAR consists of a desktop review of documents submitted by the applicant, site visits to confirm consistency between the SAR and the site condition, and independent calculations for verification.

The need for a site visit can be decided for individual hazards and is considered in the sections of this Safety Report that address external hazards.

The desktop review needs to concentrate on the soundness of the analysis and the compliance of the approaches used with those provided in the relevant national regulations and IAEA Safety Standards.

One important point to check is the linkage of the results obtained in chapter 2 ('Site characteristics') of the SAR with other chapters of the SAR to ensure consistency. For this reason, a table needs to be prepared by the applicant that shows all design basis and beyond design basis external hazard values and a reference to the chapter and Section of the SAR where these values are used.

Modelling will be the basis for the hazard analysis of some external events. Models generally contain subjective elements, and therefore the appropriate representation of epistemic uncertainties is very important.

Computer programs used in hazard evaluation need to be checked to confirm that they are properly validated and verified. Independent checks of some hazard calculations may also be needed particularly when the results, even after the necessary sensitivity and/or uncertainly analysis, indicate that acceptance criteria are met with only small margins.

4.2.4. Evaluation findings

The reviewer's evaluation of site characteristics, in accordance with the relevant regulatory criteria, needs to be documented in the SER, noting any findings that diverge from these criteria. The first step of the review is a decision on the overall quality and completeness of this Section of the SAR. It then needs to be verified that there are no issues related to site suitability.

The documentation of the review needs to include the evaluation of site characteristics with respect to the relevant regulatory criteria, including any prescribed distances and zones. The reviewer is responsible for independently verifying that the contents of the SAR comply with the acceptance criteria and for ensuring that the conclusions of the evaluation report also meet the acceptance criteria. The procedure followed by the reviewer in the evaluation needs to be briefly described.

The SER needs to include short descriptions of the site area and the site vicinity. If an exclusion area and/or a low population zone is identified, any important infrastructure in these areas needs to be indicated, and any prominent topographical features need to be highlighted.

It is necessary to verify that the applicant has provided sufficient information and that the review and calculations (if applicable) support the conclusions of the SER.

4.2.5. Implementation

The requirements established in SSR-1 [2] and the recommendations, provided in SSG-35 [10], SSG-21 [11], SSG-9 (Rev. 1) [12], SSG-18 [13], SSG-79 [18], NS-G-3.6 [19] and SSG-68 [20], need to be used in the evaluations performed by the applicant, together with all applicable national regulations. If national safety targets and acceptance criteria related to external event hazard analyses are complied with using different approaches from those provided in the referenced IAEA Safety Standards Series publications, the methods proposed by the applicant need to be duly scrutinized.

4.2.6. Detailed review of technical content

The main requirements for the evaluation of site specific hazards are Requirements 15–24 of SSR-1 [2]. Requirement 15 of SSR-1 [2] states:

"**Geological faults larger than a certain size and within a certain distance of the site and that are significant to safety shall be evaluated to identify whether these faults are to be considered capable faults. For capable faults, potential challenges to the safety of the nuclear installation in terms of ground motion and/or fault displacement hazards shall be evaluated.**"

Requirement 16 of SSR-1 [2] states:

"**An evaluation of ground motion hazards shall be conducted to provide the input needed for the seismic design or safety upgrading of the structures, systems and components of the nuclear installation, as well as the input for performing the deterministic and/or probabilistic safety analyses necessary during the lifetime of the nuclear installation.**"

Requirement 17 of SSR-1 [2] states that "**Hazards due to volcanic activity that have the potential to affect the safety of the nuclear installation shall be evaluated.**"

Requirement 18 of SSR-1 [2] states that "**Extreme meteorological hazards and their possible combinations that have the potential to affect the safety of the nuclear installation shall be evaluated.**"

Requirement 19 of SSR-1 [2] states that "**The potential for the occurrence of rare meteorological events [footnote omitted] such as lightning, tornadoes and cyclones, including information on their severity and frequency, shall be evaluated.**"

Requirement 20 of SSR-1 [2] states that "**Hazards due to flooding, considering natural and human induced events including their possible combinations, shall be evaluated.**"

Requirement 21 of SSR-1 [2] states that "**The geotechnical characteristics and geological features of subsurface materials shall be investigated, and a** soil and rock profile for the site that considers the variability and uncertainty in subsurface materials shall be derived."

Requirement 22 of SSR-1 [2] states that "**Geotechnical hazards and geological hazards, including slope instability, collapse, subsidence or uplift, and soil liquefaction, and their effect on the safety of the nuclear installation, shall be evaluated.**"

Requirement 23 of SSR-1 [2] states that "**Other natural phenomena that are specific to the region and which have the potential to affect the safety of the nuclear installation shall be investigated.**"

Requirement 24 of SSR-1 [2] states that "**The hazards associated with human induced events on the site or in the region shall be evaluated.**"

4.2.6.1. *Review of the database, models and management of uncertainties*

The database for this Section of the SAR includes the data needed for the screening out of natural or human induced external hazards from further consideration. The database needed for the evaluation of each individual hazard is discussed in Sections 4.3–4.7 of this Safety Report.

Ground motion hazards due to earthquakes, flooding hazards (e.g. coastal, river, precipitation related) and extreme meteorological hazards will always be present and need to be evaluated. For all other hazards, it is necessary to decide whether they need to be considered in the design of the nuclear installation. For these hazards, existing information is generally used in the database.

Once the database is established for each of the specific hazards, a suitable model of screening needs to be selected. As a result of the screening, the external hazard under investigation may be considered a credible hazard for the site, and consequently a detailed evaluation of the hazard is undertaken.

Accounting for and eventually quantifying both aleatory and epistemic uncertainties is essential in the screening of the external hazards. For this reason, enveloping assumptions are made when calculating screening distances. Similarly, if screening is done on the basis of probabilities, then exceedance probability values of about two to three orders of magnitude less than those corresponding to design levels are used for screening purposes.

4.2.6.2. *Fault capability*

Geological and tectonic maps of 1:25 000 to 1:50 000 scales are needed for the near region (about 25 km radius) [10]. Depending on the available database in the country, the tectonic map may already indicate capable faults. If this is not the case, the tectonic map needs to be used together with geological maps in which the ages of different formations are provided.

In general, screening is based on distance. The requirement of the absence of capable faults in the site vicinity may be taken as the basis for this screening process. A distance of 5 km may be taken as the screening distance.

The uncertainties involve several aspects. First, for many countries a 'capable fault' map (using the definition from SSR-1 [2]) might not exist. In such a case, assumptions would need to be made about the faults delineated on the tectonic map. Second, the distance as measured on a two dimensional map might not reflect the true distance, as faults have dip angles associated with them that could make them closer or farther away from the site at depth.

4.2.6.3. *Evaluation of ground motion hazards*

Regardless of the characteristics of the site, ground motion is never screened out, and consequently it is always evaluated. Since ground motion hazard evaluations have been performed for most nuclear installations worldwide, it is a mature subject. Therefore, there is consensus in greater detail on this subject than for other external hazards. This detail is reflected in SSG-9 (Rev. 1) [12].

Recommendations for the compilation of the database on ground motion hazards are well structured. For the geological, geophysical and geotechnical database, there are four spatial scales to consider. For the seismological database, the scaling is temporal and extends from paleo-seismological work to the recording of small local earthquakes.

More details on the review of this subject are provided in Section 4.7 of this Safety Report.

4.2.6.4. *Volcanic hazards*

Volcanic hazards are considered for different aspects of the safety assessment, which in turn require consideration at different spatial scales. The first concerns the decision on site acceptability. For this aspect, a near regional geological map is again needed in which all types of volcano are indicated. The exclusionary attributes of volcanoes include ground deformation, pyroclastic flows and lava flows. The second aspect is related to design and generally involves issues such as the potential for ash fall. For this aspect, a geological map of a much larger area (~500 km radius) would be necessary [10].

For volcanoes that have calderas, the screening uses distance. Geological maps would show the extent of lava flows and pyroclastic flows. Worldwide data on similar volcanoes need to be used to augment site specific data. The extent of lava flows and pyroclastic flows would generally not exceed some tens of kilometres. If the volcanic activity is due to a monogenetic volcanic field, however, and the prospective site is in close proximity to this field, a probabilistic approach would also be needed for screening out the hazard.

For volcanoes far away, where ash fall may be a concern, modelling would involve the amount and type of ash from the particular volcano as well as meteorological conditions to transport the ash from the volcano to the site.

Uncertainties in the database may be mostly related to the age dating of the volcanic activity. Uncertainties in age dating of volcanic activity would mainly affect the probability screening.

4.2.6.5. *Rare meteorological events*

Tropical storms and tornadoes are the main hazards to consider for these meteorological events. A regional (~300 km radius) historical database is needed to identify these hazards [10]. Information from neighbouring regions is also necessary to take into account that climate changes may cause shifts in the regions that may be influenced by these hazards.

The parameters needed to define tornado loads include the wind speed and the size of the tornado. The characterization of tornado loads is generally described using the Enhanced Fujita Scale. Although very strong tornadoes occur only in some parts of the world (e.g. the tornado corridor of the United States of America), smaller tornadoes may occur much more widely. Waterspouts also need to be considered, as they may affect the safety of coastal structures of the nuclear installation.

Depending on the geographical location, it may be possible to screen out one or more rare meteorological events, or at least some categories of events. As very strong tornadoes tend to occur in known areas, for example, these may be screened out if the site is located in a completely different region of the world, and if the historical database confirms that these events do not tend to occur in that region. In some cases, smaller rare meteorological events may occur in certain parts of the world. For example, in the Mediterranean area (especially in the western part) a phenomenon called 'medicane' occurs. These events are smaller than the large hurricanes of the Atlantic and the typhoons of the Pacific, but they generate significant winds and accompanying coastal flooding, which need to be considered.

Tornadoes and tropical cyclones are likely to be affected by climate change, both in terms of location and size. This introduces an additional uncertainty, as the frequency, intensity and special distribution of these events may change over time. Therefore, the reviewer needs to check that the screening process has been applied and that it has taken into account all uncertainties and potential time-dependant changes in the recurrence process.

4.2.6.6. Flooding hazards

For coastal sites, flooding can be due to storm surges, seiches, wave set-up, tsunamis or some combinations of these events. The database needed for these hazards would include pre-historical and historical data, data on offshore faults or landslides (for tsunamis), and bathymetry at regional and near regional scales.

For river sites, floods may be caused by extreme precipitation, snowmelt or upstream dam breaks. Data are needed both on the pre-historical and historical floods and on the characteristics of the river basin, as well as the physical characteristics of the river cross-section along the entire course of the river. Detailed maps of the basin (1:25 000 to 1:50 000) are needed [10].

The screening of some flood hazards may be performed if the installation grade (for all structures, systems and components important to safety) is significantly higher than the source of cooling water for the nuclear installation.

When assessing the potential for tsunami hazards at a nuclear installation site, specific guidance can help determine whether further investigations are necessary. For example, para. 5.44 of SSG-18 [13] provides a clear criterion for screening tsunami risks. It states that no additional investigations or studies are needed if the site demonstrates no evidence of tsunami occurrences and meets certain geographical or topographical conditions. These conditions include being located either more than 10 km from a sea or ocean shoreline (or 1 km from a lake or fjord shoreline) or at an elevation exceeding 50 m above the mean water level. This guidance offers a practical approach to excluding tsunami hazards under clearly defined circumstances.

Even if coastal and river flooding may be screened out due to the high installation grade, the potential for flash flooding caused by extreme precipitation cannot be ruled out and needs to be evaluated.

Modelling is not normally performed at the screening stage. Review of the modelling aspects and the management of uncertainties are discussed in Section 4.5 of this Safety Report.

4.2.6.7. Geotechnical and geological hazards

Geotechnical and geological hazards involve slope instability (landslide or rock fall), subsidence, liquefaction and collapse (due to the existence of cavities). These hazards are localized, and they affect the safety of the nuclear installation only if they are within the site vicinity (~5 km radius) or the site area. Existing geological, hydrogeological and geophysical information presented on maps of 1:25 000 scale is needed to identify the potential for these hazards. It might not be possible to rule out these hazards completely until the levels of the vibratory ground motion have been assessed, as this motion may activate the geotechnical hazard [10]. Therefore, the necessary database might not be complete until the seismic hazards have been evaluated. Geotechnical hazards might or might not be initiated by vibratory ground motion. For example, slope failures may also be initiated by excessive precipitation.

Screening out geotechnical hazards might not be straightforward because the exclusionary attribute of a geotechnical condition would need to be considered together with the feasibility of remedial engineering measures. Furthermore, the geotechnical data that form the basis for screening decisions in the site selection process are generally not complete. For this reason, decisions regarding the screening out of geotechnical and geological hazards need to be revisited as detailed site investigations progress.

Modelling is not normally performed at the screening stage. Review of the modelling aspects and the management of uncertainties is discussed in Section 4.7 of this Safety Report.

4.2.6.8. Other natural hazards

Some natural hazards are very much region specific and depend mainly on climate. In arid areas, dust storms and sandstorms may occur. In very cold climates, permafrost and icing of the cooling water source may be sources of hazard. Extreme temperatures of the air and the water are always sources of concern. In the screening of these hazards, the effects of climate change need to be considered.

Similar to rare meteorological events, a regional (~300 km radius) historical database is needed to identify these hazards. Information from neighbouring regions is also necessary, as historical databases rarely extend geographically to a statistically meaningful extent [10]. The data from neighbouring (and climatically similar) regions allows ergodic assumptions to be applied to these climate-dependent hazard processes.

There may also be other types of hazard specific to the region. One example is the possible existence of mud volcanoes. They are caused by gases under pressure, which may cause small eruptions, in turn, to ground deformations associated with mud volcano activity, which may be expressed as mounds rising tens of metres. Geological and geophysical data at near regional scale are needed to identify these deformations. The hazard from mud volcanoes may be screened out using both screening distance and screening probability.

Modelling of sandstorms and dust storms will uses geotechnical data to simulate the grain size distribution and meteorological data for wind speed and direction.

Extreme air or water temperature may be modelled through statistical distributions, but constraining these models is often found to be difficult. Data on extreme air or water temperatures observed in neighbouring regions with similar climatic characteristics may be used to constrain the statistical models, under the assumption of ergodicity.

Climate change is always a source of major uncertainty. Several climate change models may need to be used to address epistemic uncertainties.

4.2.6.9. *Hazards due to human induced events*

Most human induced hazards may have high intensity but short range. In general, the maximum extent for the region of interest would be within a 10–15 km radius. The first step in the screening process for human induced events is the source display map (~1:25 000 scale), which identifies all potential sources of human induced events that may impact nuclear safety [10]. Normally, it is not difficult to get this information from local and national official sources. However, some sources may have confidentiality issues to deal with. Military installations, for example, would not readily disclose the arsenal they possess, nor would they provide information regarding the activities going on at the installation. Also, it would be hard to estimate both the type and amount of hazardous material that may be transported in ships, trains and trucks.

Pipelines may initiate several types of hazard, including explosions leading to pressure waves and gas clouds that might result in deflagration or, under very unfavourable conditions, detonation. Missiles may also result from such explosions.

In addition to these uncertainties in estimating human induced hazards, it needs to be remembered that these hazards are highly non-stationary; that is, their variety and amount tend to increase with time.

Modelling is normally not performed at the screening stage. Review of the modelling aspects and the management of uncertainties are discussed in Section 4.3 of this Safety Report.

4.2.7. Reviews for small modular reactors

4.2.7.1. *Fault capability*

As fault capability is a potentially exclusionary external hazard, no grading of the requirements or recommendations is envisaged. It is unlikely that the differences in the size of the footprint of an SMR in comparison to a large nuclear power plant would make a significant difference in screening distances. Therefore, an approach similar to that used for large nuclear power plants needs to be followed for the screening process (see Section 4.2.6 of this Safety Report).

4.2.7.2. *Evaluation of ground motion hazards*

The ground motion hazard is always evaluated, and there is no screening process. Details of how a seismic hazard evaluation programme may be graded are discussed in Section 4.7 of this Safety Report.

4.2.7.3. *Volcanic hazards*

Exclusionary aspects of volcano hazards will be treated in a similar way to the requirements and recommendations for large nuclear power plants. It can be assumed that an SMR will have a smaller footprint than a large nuclear power plant; however, a large nuclear power plant may have more structural robustness (e.g. for impact and blast loads). Therefore, screening distance values related to some effects of volcanic hazards may be significantly larger than those for a large nuclear power plant. That the SMR footprint is smaller than the footprint of a large nuclear power plant might not produce a significant benefit in the probabilistic screening, as the difference in the footprint is likely to be less than an order of magnitude. Therefore, an approach similar to that used for large nuclear power plants needs to be followed for the screening process (see Section 4.2.6 of this Safety Report).

4.2.7.4. *Rare meteorological events*

That the SMR footprint is smaller than the footprint of a large nuclear power plant might not bring a significant benefit in probabilistic screening for tropical cyclones, typhoons, hurricanes, tornadoes and waterspouts, as the difference in the footprint is likely to be less than an order of magnitude.

The screening process for rare meteorological events is discussed in Section 4.2.6 of this Safety Report.

4.2.7.5. *Flooding hazards*

The screening process for flood hazards for SMRs is not different from the process applied for large nuclear power plants. Therefore, the practical guidance provided in Section 4.2.6 of this Safety Report is applicable.

4.2.7.6. *Geotechnical and geological hazards*

That the SMR footprint is smaller than the footprint of a large nuclear power plant might not produce a significant benefit in the screening in relation to geotechnical or geological hazards, as the difference in the footprint is likely to be less than an order of magnitude. Therefore, an approach similar to that used for large nuclear power plants needs to be followed for the screening process (see Section 4.2.6 of this Safety Report).

4.2.7.7. Other natural hazards

In relation to external hazards such as sandstorms, dust storms and mud volcanoes, there would be no difference in the approach for SMRs compared with the approach for large nuclear power plants (see Section 4.2.6).

Screening for extreme temperatures may be different depending on the design of the SMR and especially its UHS. In any case, uncertainties may be considerable and need to be accounted for in the screening process.

4.2.7.8. Hazards due to human induced events

Two types of screening are generally performed for human induced events; the first is based on distance, and the second is based on probability. The screening distance value for large nuclear power plants is set for each hazard from each source with respect to the design of the nuclear power plant for loads such as impact, blast, thermal and vibration. It is expected that, in general, SMRs might not be as robust as large nuclear power plants with respect to their design against these loads, which may lead to larger screening distance values.

However, the exposed profile of SMR structures may be substantially smaller than that of a large nuclear power plant. This difference would be more pronounced if the parts of an SMR that are important to safety are significantly embedded. This is a consideration for the possibility of screening out of some missile impact scenarios, such as aircraft crashes.

4.2.8. References for the review

The references for Section 4.2 include SSG-61 [1], SSR-1 [2], SSG-35 [10], SSG-21 [11], SSG-9 (Rev. 1) [12], SSG-18 [13], SSG-79 [18], NS-G-3.6 [19] and SSG-68 [20].

4.3. PROXIMITY OF INDUSTRIAL, TRANSPORTATION AND OTHER FACILITIES

4.3.1. Areas of review

The target of the SAR review in Section 4.3 includes the information recommended in paras 3.2.18 and 3.2.19 of SSG-61 [1].

The basic information to be reviewed is the source display map that provides all potential sources of human induced events in the site vicinity and the near region of the nuclear installation.

For each source, relevant events that may affect the safety of the nuclear installation are postulated. Conservative estimates need to be used, and these estimates need to also consider future development throughout the lifetime of the nuclear installation. New projects (e.g. roads, airports, storage facilities) need to be identified, together with the associated hazards.

The review checks the consistency of the screening values with those identified and substantiated in the previous Section of this Safety Report.

Parameterization and enveloping are very important parts of screening. (For more detailed information, refer to all of the tables in the appendix and annex of SSG-79 [18]). Enveloping involves the screening out of some hazards because their effects are bounded by hazards already considered for analysis. However, before a hazard can be screened out, it needs to be confirmed that all effects associated with that hazard are covered by other hazards that have been considered for analysis. For this reason, hazards need to be parameterized. For example, a railroad accident may result in impact loads due to missiles and blast loads due to the explosion. For this event to be enveloped, both effects (i.e. parameters related to impact and blast) need to be enveloped by other human induced events.

Any security related design features or administrative measures are identified and coordinated with those related to accidental human induced hazards. In particular, if the nuclear installation has been designed to withstand the crash of a large commercial airplane, parameters related to the selected scenarios can be used to screen out some accidental scenarios, keeping in mind that a large commercial aircraft crash is generally considered a beyond design basis event.

Protection against human induced external events is provided through design, site protection features or administrative measures. The latter is the least preferred and needs to be scrutinized in the review. It is preferable that administrative measures are used as an extra layer of protection and not as the only means of protection.

Human induced hazards are non-stationary in nature, and both their frequency and their size may significantly increase during the lifetime of the nuclear installation.

Requirement 24 of SSR-1 [2] is the most relevant for these areas of the review. Section 4.3 interfaces with Sections 4.1, 4.2, 4.4, 4.10 and 4.11 of this Safety Report.

The paragraphs in SSG-61 [1] relevant to these areas of the review state:

"3.2.18. This Section should describe the locations and transport routes that represent potential risks for the plant and the results of a detailed evaluation of the effects of potential accidents at industrial, transportation or other facilities in the vicinity of the site. Projected developments in the vicinity over the envisaged lifetime of the nuclear power plant relating to this information should also be presented and updated, as required, in future stages of the safety analysis report.

"3.2.19. Any identified risks considered in determining the design basis should be included to help determine whether any additional measures are necessary to mitigate the adverse effects of potential incidents."

4.3.2. Acceptance criteria

The annual probability of exceedance to be assumed for external hazards needs to be prescribed in the national regulation. If they have not been prescribed in the national regulation, relevant recommendations from IAEA Safety Guides may be used, if available. When these values are not provided in the national regulation, it is essential that the values are discussed and agreed with the regulatory body.

4.3.3. Review procedure

A typical safety review of an SAR consists of a desktop review of documents submitted by the applicant, site visits to confirm consistency between the SAR and the site condition, and independent calculations for verification.

The review is performed mainly using documentation. Source display maps drawn to appropriate scales need to be prepared by the applicant. In particular, the geographical coordinates of key points, such as the locations of potential human induced hazard sources, need to be provided.

The site area for the installation needs to be defined in accordance with SSR-1 [2]. This is the area under the administrative control of the operating organization. If there is more than one collocated nuclear installation under the administrative control of different organizations, these are considered potential sources of human induced hazard.

A site visit is necessary in order to understand the setting of the site with respect to the near region and site vicinity infrastructure, including all of the sources of human induced hazards.

Computer programs used in hazard evaluation need to be checked to confirm that they are properly validated and verified. Independent checks of some hazard calculations may also be needed particularly when the results, even after the necessary sensitivity and/or uncertainly analysis, indicate that acceptance criteria are met with only small margins.

4.3.4. Evaluation findings

The reviewer's evaluation of site characteristics, in accordance with the relevant regulatory criteria, needs to be documented in the SER, noting any findings that diverge from these criteria. The first step of the review is a decision on the overall quality and completeness of this Section of the SAR. It then needs to be verified that there are no issues related to site suitability.

The documentation of the review needs to include the evaluation of human induced hazards with respect to the relevant regulatory criteria. The reviewer is responsible for independently verifying that the contents of the SAR comply with the acceptance criteria and for ensuring that the conclusions of the evaluation report also meet the acceptance criteria. The procedure followed by the reviewer in the evaluation needs to be briefly described. If needed, the evaluation may include the performance of independent calculations to ascertain compliance.

The SER needs to include short descriptions of the approaches adopted for human induced external hazard analyses, including the database used and the results obtained. This includes the screening methods and values, as well as the target probabilities, used in the design bases and beyond design bases for human induced hazards.

It is necessary to verify that the applicant has provided sufficient information and that the review and calculations (if applicable) support the conclusions of the SER.

4.3.5. Implementation

The requirements established in SSR-1 [2] and the recommendations provided in SSG-79 [18] need to be used in the evaluations performed by the applicant. If national safety targets and acceptance criteria related to human induced hazards are complied using different approaches from those set forth in IAEA Safety Standards Series publications, the methods proposed by the applicant need to be duly scrutinized.

4.3.6. Detailed review of technical content

The main requirement regarding the proximity of industrial, transportation and other facilities is given in SSR-1 [2]. Requirement 24 and associated paras 5.33–5.37 of SSR-1 [2] state:

"The hazards associated with human induced events on the site or in the region shall be evaluated.

"5.33. Human induced events to be addressed shall include, but shall not be limited to:

(a) Events associated with nearby land, river, sea or air transport (e.g. collisions and explosions);
(b) Fire, explosions, missile generation and releases of hazardous gases from industrial facilities near the site;
(c) Electromagnetic interference.

"5.34.   Human activities that might influence the type or severity of natural hazards, such as resource extraction or other significant re-contouring of land or water or reservoir induced seismicity, shall be considered.

"Aircraft crashes

"5.35. The potential for accidental aircraft crashes on the site shall be assessed with account taken, to the extent practicable, of potential changes in future air traffic and aircraft characteristics.

"Chemical hazards

"5.36. Current or foreseeable activities in the region surrounding the site that involve the handling, processing, transport and/or storage of chemicals having a potential for explosions or for producing gas clouds capable of deflagration or detonation shall be addressed.

"5.37. Hazards associated with chemical explosions or other releases shall be expressed in terms of heat, overpressure and toxicity (if applicable), with account taken of the effect of distance and non-favourable combinations of atmospheric conditions at the site. In addition, the potential effects of such events on site workers shall be evaluated."

4.3.6.1. *Hazards due to transportation*

Hazards due to transportation include railway, highway, waterway and airway traffic. Data related to these modes of transportation need to be collected. Pipelines may initiate multiple hazards such as pressure waves, deflagration, detonation and/or missiles [18].

For railways, the important information is related to the location and the passage frequency — data on both of which are generally easy to obtain — as well as the nature and amount of transported cargo, for which data are harder to obtain, especially if the route is international. For highways, the information needed is similar, although the variability in the nature of transported cargo is much more than for railways.

If the site is coastal, then the proximity of ports and harbours is very important, and shipping lanes need to be identified. The bathymetry of the site vicinity is also important to understand the closest proximity that vessels might reach if they went adrift in poor weather conditions. Ships have a very large capacity for carrying hazardous material, and it is difficult to predict the cargo of ships well in advance.

If the site is on a river, barge traffic on the river may pose a hazard. The proximity of these barges to the site is necessarily very close, and barges can carry large amounts of hazardous material.

The most important data to be collected related to air traffic are (i) the location of airfields in the near region, (ii) the air traffic at those airfields, (iii) the types of aircraft involved in the air traffic in the near region, (iv) the location of air corridors, and (v) the types and frequency of aircraft using these corridors. It is, of course, important to distinguish differences between civilian and military aircraft as their sizes, speeds and accident statistics.

For railway and highway traffic, the hazard would be related to explosions (which may also include missiles) and releases of hazardous gases. The standard procedures suggested in SSG-79 [18] need to be used to screen by distance and probability. For those parameters that have not been screened out, the design basis needs to be established.

For waterway traffic, there is an additional consideration that involves the possibility of direct impact by a vessel with the coastal structures of the nuclear installation. The screening of such an event needs to be based on probability. In other respects, the modelling and analysis are similar to those for railway and highway traffic.

Uncertainties related to transportation are very significant. The data collected are related to the present conditions. However, there is a need to extrapolate these data up to approximately 80 years in the future, depending on the projected lifetime of the installation. The uncertainties involved are as follows:

(a) Uncertainties related to the economic and industrial development of the region. Generally, governmental plans cover the near future, and it will therefore not be possible for some major development plans that may be proposed in the more distant future to be considered in the modelling.

(b) Types and capacities of different transportation vehicles that may be developed in the future.

(c) The considerable variability in the proximity of future transportation routes to the site as well as the frequency of traffic, even if industrial and economic development plans are well estimated.

4.3.6.2. *Aircraft crashes*

The treatment of aircraft crashes has become more deterministic and prescriptive. Therefore, if there is a regulatory requirement for an aircraft crash scenario (or more than one scenario), then the screening process for an accidental (i.e. probability based) aircraft crash needs to be checked to verify that it is bounded by the deterministic scenario. This bounding of the accidental crash by the deterministic scenario may facilitate the screening process for the accidental crash. However, if there is only one deterministic scenario, that scenario may be associated with design extension conditions.

Considerations for the uncertainties related to an accidental aircraft crash are the same as discussed in Section 4.3.6.1 of this Safety Report. The reviewer needs to verify that the margins related to the associated design bases are sufficient to consider these uncertainties.

4.3.6.3. *Fire, explosions, missile generation and releases of hazardous gases from industrial facilities near the site*

Data related to stationary sources in the site vicinity are documented on the source display map at a sufficiently large scale (~1:5000 to 1:10 000). Stationary sources may be civilian or military [10]. They may involve petrochemical or hazardous material storage facilities, arsenals and shooting ranges. Data need to be collected on the size of the hazard (e.g. the volume of explosive material) and its frequency. Screening by distance is performed using the size of the hazard. This information might not be readily available for military facilities, and conservative assumptions may need to be made.

Information on accident frequency is obtained from worldwide generic databases for similar facilities. Partly due to the uncertainties involved, the screening frequency levels are taken to be two to three orders of magnitude smaller than what was used as the design basis.

Similar to transportation facilities, stationary sources also tend to change with time. Furthermore, military weapons are likely to get larger in size and longer in range. It is also possible that military sites may store hazardous materials that are not publicly known to civilians.

The reviewer needs to check that all of these uncertainties have been considered in the evaluation of human induced hazards.

4.3.6.4. *Electromagnetic interference*

Paragraph 10.9 of SSG-79 [18] indicates that electromagnetic interference can affect the functionality of electronic devices and that it can be initiated by both on-site sources (e.g. high voltage switchgear, portable telephones, portable electronic devices, computers) and off-site sources (e.g. radio interference, the telephone network). Also, the presence of central telephone installations close to the site could give rise to interference.

It is recommended in SSG-79 [18] that, during the site evaluation stage, potential sources of interference need to be identified and quantified (e.g. their intensity and frequency) and need to be monitored over the lifetime of the nuclear installation for the purpose of ensuring the proper qualification of components.

The region from which data are collected needs to be large enough that all potential interference can be identified.

4.3.7. Reviews for small modular reactors

4.3.7.1. *Hazards due to transportation*

Hazards that originate on railways, highways and sea routes need to be treated in a similar way to that described in Section 4.3.6 of this Safety Report, albeit possibly using larger screening distances. This is because, typically, SMRs would be expected to be less robust against loads such as blast and impact because their wall thickness would be less than that of large nuclear power plants. Due to the smaller size of SMRs, the probability of a missile impacting an SMR would be lower than for a large nuclear power plant, especially if the SMR is well embedded, which is a design preferred by some SMR designers.

The scenario of a large commercial aircraft crashing into a large nuclear power plant also needs to be carefully evaluated for SMRs. For most sites, an accidental crash of this kind can be screened out on a probability basis. However, the scenario is generally a deterministic one, prescribed by the regulatory body. One approach for demonstrating the protection of the SMR may be to demonstrate that the small profile of the part of the installation that is important to safety makes direct impact by an aircraft crash almost impossible, even in the case of a deliberate attack. For some SMRs, this might not be the case. One example would be the marine based SMRs where the installation is exposed; the impact of a large aircraft on these SMRs would potentially result in unacceptable consequences.

4.3.7.2. Fire, explosions, missile generation and releases of hazardous gases from industrial facilities near the site

For missiles, the discussion in Section 4.3.7.1 of this Safety Report will also apply. For all other types of human induced hazard, the discussion in Section 4.3.6 of this Safety Report will apply.

4.3.7.3. Electromagnetic interference

In general, the discussion in Section 4.3.6 of this Safety Report will apply.

4.3.8. References for the review

The references for Section 4.3 include SSG-61 [1], SSR-1 [2], SSG-35 [10] and SSG-79 [18].

4.4. ACTIVITIES AT THE SITE THAT MIGHT INFLUENCE THE SAFETY OF THE NUCLEAR INSTALLATION

4.4.1. Areas of review

The target of the SAR review in Section 4.4 includes the information recommended in paras 3.2.20 and 3.2.21 of SSG-61 [1].

The number of administrative units at the site first needs to be clarified. If there are multiple nuclear installations collocated under more than one administrative organization, then coordination is needed, and the procedures for this coordination need to be reviewed. Section 4.4 is related only to the potential hazards under the administrative control of the operating organization. Human induced hazards related to collocated installations under a different administrative control are considered in Section 4.3 of this Safety Report.

A source map of the site area (including transport routes, structures containing explosive or flammable substances, radioactive material, protective structures, and site protection features such as dykes) may be reviewed, together with procedures for recording any changes. An inspection and maintenance programme needs to be implemented for all site protection features.

Requirement 24 and, if site protection measures are needed, Requirement 8 of SSR-1 [2] are the most relevant for these areas of the review.

Section 4.4 interfaces with Sections 4.1–4.3, 4.10 and 4.11 of this Safety Report.

The paragraphs in SSG-61 [1] relevant to these areas of the review state:

"3.2.20. In this section, any processes or activities at the site that, if incorrectly carried out, could affect or influence the safe operation of the plant should be presented and described. Examples of such processes or activities include vehicular transport in the plant area; storage of fuels, gases and other chemicals; and activities potentially leading to intakes of or contamination by harmful particles, smoke or gases (e.g. intakes of air through ventilation systems).

"3.2.21. Measures for site protection (e.g. dams or dykes for flood control and drainage) and any modifications to the site (e.g. soil substitution, modifications to the site elevation) are usually considered at the site characterization stage, and their assessment in relation to the design basis should be included in this Section of the safety analysis report."

4.4.2. Acceptance criteria

Acceptance criteria for on-site human induced events are not different from those outside the fence. The annual probability of exceedance to be assumed for external hazards needs to be prescribed in the national regulations. If they have not been prescribed in the national regulation, relevant recommendations mentioned in IAEA Safety Guides may be used, if available. When these values are not provided in the national regulation, it is essential that the values be discussed and agreed with the regulatory body.

4.4.3. Review procedure

A typical safety review of an SAR consists of a desktop review of documents submitted by the applicant, site visits to confirm consistency between the SAR and the site condition, and independent calculations for verification.

The review is mainly performed from documents. A source display map drawn to appropriate scale needs to be prepared by the operating organization.

In particular, key points, such as the locations of potential human induced hazard sources, are to be provided on the site plan.

A site visit is necessary in order to understand the installation layout and the respective locations of potential sources of human induced hazards.

Independent checks of some hazard calculations may also be needed particularly when the results, even after the necessary sensitivity and/or uncertainly analysis, indicate that acceptance criteria are met with only small margins.

Structures that are used for site protection purposes (e.g. barriers, dykes) need to be indicated on the site layout plan.

If administrative measures are used for protection against human induced hazards, procedures related to these measures are to be reviewed.

4.4.4. Evaluation findings

The reviewer's evaluation of site characteristics, in accordance with the relevant regulatory criteria, needs to be documented in the SER, noting any findings that diverge from these criteria. The first step of the review is a decision on the overall quality and completeness of this Section of the SAR.

The documentation of the review needs to include the evaluation of human induced hazards (within the area under the administrative control of the operating organization) with respect to the relevant regulatory criteria. The reviewer is responsible for independently verifying that the contents of the SAR comply with the acceptance criteria and for ensuring that the conclusions of the evaluation report also meet the acceptance criteria. The procedure followed in the evaluation needs to be briefly described. If needed, the evaluation may include the performance of independent calculations to ascertain compliance.

The SER needs to include short descriptions of the approaches adopted for human induced external hazard evaluation, including the database used and the results obtained. This includes the screening methods and values, as well as the target probabilities, used in the design basis and beyond design basis for human induced hazards.

Structures that are used for site protection purposes (e.g. barriers, dykes) need to be reviewed for their adequacy in performing their functions.

Procedures supporting any administrative measures used for protection against human induced hazards need to be reviewed.

It is necessary to verify that the applicant has provided sufficient information and that the review and calculations (if applicable) support the conclusions of the SER.

4.4.5. Implementation

The requirements established in SSR-1 [2] and the recommendations provided in SSG-79 [18] need to be used in the evaluations performed by the applicant. If national safety targets and acceptance criteria related to human induced hazards are complied with using different approaches from those provided in IAEA Safety Standards Series publications, the methods proposed by the applicant need to be duly scrutinized.

4.4.6. Detailed review of technical content

Requirement 24 of SSR-1 [2] states that **"The hazards associated with human induced events on the site or in the region shall be evaluated."**

4.4.6.1. Review of the database, the models and management of uncertainties

Human induced events that occur at the site include events associated with the installation under review and managed by the administration in charge of preparing the SAR, as well as any other collocated nuclear installations with different operating organizations. The data requirements, as well as the models used for hazard evaluation, will vary for these two cases.

The data requirements for the nuclear installation under review involve a source display map that indicates all potential stationary and mobile hazards. The latter would be related to delivery vehicles such as trucks that carry hazardous material. All relevant parameters to be used in hazard analysis and associated with the sources on the source display map need to be considered. These involve the following:

(a) Substances that may be released and affect heating, ventilation and air conditioning systems;
(b) Explosives that may generate blast and/or impact loads;
(c) Substances that may start fires;
(d) Electromagnetic interference.

The reviewer needs to confirm that this information is appropriately documented and updated at regular intervals.

The information on the on-site sources is expected to be much more reliable than the data collected for human induced events that occur outside of the site. However, it needs to be remembered that, as these sources are very close to the structures, systems and components of the nuclear installation, their capability for causing failures is significant. Therefore, the modelling and hazard analysis needs to be performed using appropriate margins.

For collocated nuclear installations that have a different operating organization from the installation under review, there are additional hazards to consider than those listed earlier in Section 4.4. If, for example, the collocated installation was built much earlier using outdated IAEA safety standards and criteria, then the possibility of a nuclear accident at this installation would need to be considered as an external hazard for the nuclear installation under review. The essential point here is the difference between the probabilities of having an accident for the installation under review and for the other collocated installations. If there is a significant difference, then the installation with greater potential for having an accident becomes a source of external hazard for the other installation.

Therefore, the database for this aspect of the review needs to include relevant safety information from collocated installations.

Uncertainties related to the information coming from the nuclear installation under review and from installations that are collocated are expected to be significantly less than those for external hazards that originate outside the site. However, the reviewer needs to verify that sufficient margins are considered in the hazard analyses due to the proximity of these sources.

4.4.7. Reviews for small modular reactors

The discussion in Section 4.4.6 of this Safety Report also applies to SMRs.

4.4.8. References for the review

The references for Section 4.4 include SSG-61 [1], SSR-1 [2], SSG-35 [10] and SSG-79 [18].

4.5. HYDROLOGY

4.5.1. Areas of review

The target of the SAR review in Section 4.5 includes the information recommended in paras 3.2.22–3.2.24 of SSG-61 [1].

There are three distinct areas of review for hydrology. The first is related to the safety of the UHS. The second is river or coastal flooding. The third is transport and diffusion of effluents in the hydrosphere. The review of flooding needs to be based on SSG-18 [13], whereas the review of transport and the diffusion of effluents in the hydrosphere needs to refer to NS-G-3.2 [9]. The review of the safety of the UHS is based on Requirement 11 of SSR-1 [2].

4.5.1.1. *Ultimate heat sink*

The UHS of a nuclear power plant is generally a body of water that contains a sufficient amount of water to fulfil the heat removal requirements of the reactor and associated safety systems, even under the most unfavourable conditions. Therefore, all hazards that could affect the availability of the UHS need to be investigated and evaluated.

4.5.1.2. *Flooding*

Depending on the site, flooding could be coastal or from a river; a combination of these types of flooding would also be possible for sites on estuaries.

The following types of flooding hazard can be considered: storm surges, wind generated waves, tsunamis, seiches, extreme precipitation events, floods due to sudden release of impounded water, bores, and mechanically induced waves.[2]

For the evaluation of these hazards, the first important step is the development of an appropriate database. Recommendations related to the database are provided in Section 3 of SSG-18 [13]. The results from the individual hazard analyses for these flood types need to be combined with effects such as tides. The simultaneous occurrence of some of these events may also be possible and needs to be considered. The effects of climate change also need to be considered.

The eventual objective is the determination of the 'maximum flood level', which is calculated conservatively using either probabilistic or deterministic methods (or a combination of the two).

4.5.1.3. *Transport and diffusion of effluents in the hydrosphere*

Transport and diffusion of effluents in the hydrosphere refers to their transport and diffusion either in surface waters (e.g. rivers, lakes, oceans) or in groundwater. Section 3 of NS-G-3.2 [9] provides recommendations on the database and methods necessary to evaluate the transport and diffusion both in surface waters and ground waters.

---

[2] Depending on whether the site is coastal or on a river, not all of these hazards may be relevant.

4.5.1.4. Links between areas of review and IAEA safety standards

Requirements 6–8, 10, 11, 20, 21 and 25 of SSR-1 [2] are the most relevant for these areas of the review.

Section 4.5 interfaces with Sections 4.2, 4.6, 4.8, 4.10 and 4.11 of this Safety Report.

The paragraphs in SSG-61 [1] relevant to these areas of review state:

"3.2.22. This Section of the safety analysis report should present sufficient information to enable evaluation of the potential implications of hydrological conditions at the site for the plant design and safe operation, with special attention devoted to conditions that potentially affect residual heat removal to the ultimate heat sink. Cooling water channels and reservoirs to be used for cooling the plant should be described. Low water conditions and the possibility of using groundwater sources in extraordinary situations should also be considered.

"3.2.23. The conditions that should be taken into account in this Section include potential floods resulting from phenomena such as abnormal ice effects and heavy rainfall, as well as runoff floods from watercourses, reservoirs, adjacent drainage areas and site drainage. This Section should also include consideration of flood waves resulting from dam failures; flooding caused by landslides, ice jams and other ice related flooding; and seismically generated, water based effects on and off the site. For coastal and estuary sites, evaluations should include storm surge, tsunamis and seiches. For both coastal and riverine flooding, reasonable combinations of hazards (e.g. tides, strong wind) and potential effects of climate change should be considered.

"3.2.24. The information given in this Section should be prepared in such a way as to enable the assessment of (i) the transport of radionuclides in groundwater and the surface water system and (ii) the dispersion of radionuclides through the environment. This information should also include a characterization of the hydrogeological subsurface properties and surface water features to enable an assessment of the measures taken to preclude the release of radionuclides to the environment."

4.5.2. Acceptance criteria

4.5.2.1. *Ultimate heat sink*

The availability of the UHS is one of the most critical safety considerations for a nuclear power plant, and it is closely linked to the target core damage frequency of the plant. Therefore, the acceptance criteria can be based on the accident analysis or the probabilistic safety analysis of the plant.

4.5.2.2. *Flooding*

For flooding, an acceptance criterion would be a 'dry site': a site with such an elevation that the 'maximum flood level' is less than this elevation, with a conservative margin. Only in exceptional cases — and with the explicit permission of the regulatory body — may the design basis flood level exceed the installation grade and may protection be ensured by features such as watertight doors.

The annual probability of exceedance associated with the design basis flood (which is based on the 'maximum flood level') is prescribed by the regulatory body.

For river flooding, the failure of water retaining structures (e.g. upstream dams and dykes) needs to be considered. Combinations of flooding need to be considered, such as a river flood and a flash flood due to intense precipitation and/or snowmelt as well as high groundwater.

For coastal sites, there may be many sources that contribute to flooding, such as precipitation, high groundwater, wave action, storm surges, seiches and tsunamis. If the site is on an estuary, the interaction between river flooding and coastal flooding needs to be considered. The acceptance criteria when combining these effects are provided by the regulatory body.

For both coastal and river flooding, rare combinations of different sources of flooding may be considered in determining the beyond design basis flood. The protection of the nuclear installation against the beyond design basis flood may be addressed using engineering solutions, such as watertight doors. However, this solution needs to be discussed and agreed with the regulatory body.

4.5.2.3. *Transport and diffusion of effluents in the hydrosphere*

The transport and diffusion of effluents in surface waters may occur when there is a discharge directly into a river, lake or ocean from the nuclear installation. It could also occur from the fallout of gaseous effluents. The transport and diffusion of effluents in groundwater may occur when there is discharge to the ground or, again, from the fallout of gaseous effluents. Discharges during normal operation and during accident conditions need to be considered. Acceptance criteria (for both normal operation and accidental discharges) are generally not related to transport or diffusion properties.

4.5.3. Review procedure

A typical safety review of an SAR consists of a desktop review of documents submitted by the applicant, site visits to confirm consistency between the SAR and the site condition, and independent calculations for verification.

The review is mainly performed from documents. Maps drawn to appropriate scales need to be prepared by the operating organization. In particular, the locations of key points, such as potential sources of hydrological event induced hazards (e.g. rivers, sea, dykes, seawalls), need to be provided on the site plan.

A site visit is necessary in order to understand the installation layout and the respective locations of potential sources of hydrological event induced hazards.

Independent checks of some hazard calculations may also be needed particularly when the results, even after the necessary sensitivity and/or uncertainly analysis, indicate that acceptance criteria are met with only small margins.

The review of both pre-flood and post-flood procedures is also needed. This involves the review of warning systems and procedures for operator actions during and after the flood.

For the transport and diffusion of effluents, the review needs very detailed hydrogeological mapping of the site vicinity (three dimensional representation may be needed) as well as detailed bathymetry maps. Mathematical models need to be scaled to real measurements, and therefore the database for these measurements needs to be reviewed.

Structures that are used for site protection purposes (e.g. barriers, dykes, seawalls) need to be indicated on the site layout plan.

If administrative measures are used for protection against hydrological event induced hazards, the procedures related to these measures need to be reviewed.

4.5.4. Evaluation findings

The reviewer's evaluation of site characteristics, in accordance with the relevant regulatory criteria, needs to be documented in the SER, noting any findings that diverge from these criteria. The first step of the review is a decision on the overall quality and completeness of this Section of the SAR. It then needs to be verified that there are no issues related to site suitability.

The documentation of the review needs to include the evaluation of flood hazards with respect to the relevant regulatory criteria. The reviewer is responsible for independently verifying that the contents of the SAR comply with the acceptance criteria and for ensuring that the conclusions of the evaluation report also meet the acceptance criteria. The procedure followed in the evaluation needs to be briefly described. If needed, the evaluation may include the performance of independent calculations to ascertain compliance.

The SER needs to include short descriptions of the approaches adopted for flood hazard analyses, including the database used and the results obtained. This includes the target probabilities used in the design basis and beyond design basis for flood hazards.

Structures that are used for site protection purposes (e.g. dykes) need to be reviewed for their adequacy in performing their functions.

Procedures supporting any administrative measures used for protection against flood hazards need to be reviewed.

It is necessary to verify that the applicant has provided sufficient information and that the review and calculations (if applicable) support the conclusions of the SER.

4.5.5. Implementation

The requirements established in SSR-1 [2] and the recommendations provided in NS-G-3.2 [9] and SSG-18 [13] need to be used in the evaluations performed by the applicant, together with all applicable national regulations. If national safety targets and acceptance criteria related to hydrology are complied with using different approaches from those provided in the national regulations and IAEA safety standards, the methods proposed by the applicant need to be duly scrutinized.

4.5.6. Detailed review of technical content

Hydrology is associated with several potential effects on the safety of a nuclear installation: the safety of the UHS, the dispersion of radioactive effluents in the hydrosphere, and flooding.

Four requirements in SSR-1 [2] address these topics, and they will be treated separately in this Safety Report and need to be reviewed accordingly.

Requirement 11 (see Section 3.1 of this Safety Report) and associated paras 4.36 and 4.37 of SSR-1 [2] establish special considerations for the UHS for nuclear installations. Paragraphs 4.36 and 4.37 of SSR-1 [2] state:

"4.36. As appropriate for the ultimate heat sink under consideration, data for the following shall be evaluated:

(a) Air temperature and humidity;
(b) Water depth and temperature;
(c) Water quality characteristics, including turbidity, suspended solids, floating debris, and chemical and biochemical changes (both natural and human induced changes);
(d) Availability and sustainability of the water flow (for a river), minimum and maximum water level and the period of time for which safety related supplies of cooling water are at a minimum level, with account taken of the potential for failure of water control structures.

"4.37. All natural and human induced external events that could cause a loss of the ultimate heat sink shall be identified and evaluated."

The data to be collected in relation to the UHS are specified in para. 4.36 of SSR-1 [2]. Accordingly, water depth and temperature, water quality characteristics, and the availability and sustainability of water need to be reviewed. These parameters would either be part of an established monitoring programme or be checked by the operating organization on a periodic basis. Therefore, the reviewer needs to check that appropriate quality documentation for ensuring this provision has been established and is being appropriately implemented.

Uncertainties in this database are expected to be insignificant. However, extrapolations and modelling for the future would need to consider non-stationarities in the data due to climate change.

With regard to the potential impact of external hazards on the UHS (see para. 4.37 of SSR-1 [2]), data requirements, modelling and the treatment of uncertainties are addressed in Sections 4.2, 4.3 and 4.5–4.7 of this Safety Report.

Direct exposure pathways would involve the exposure of the nearby population to ionizing radiation through contact without an intermediary. Indirect exposure pathways mainly involve food chains, which may impact populations far away from the nuclear installation.

Requirement 12 (see Section 3.1 of this Safety Report) and associated paras 4.38–4.40 of SSR-1 [2] relates to the potential effects of the nuclear installation on people and the environment. Paragraphs 4.38–4.40 of SSR-1 [2] state:

"4.38. The potential effects of the nuclear installation on people and the environment shall be estimated by considering the postulated accident scenarios (including the resulting source terms) and taking into account the feasibility of planning effective emergency response actions at the site and in the external zone. These estimates shall be confirmed when the design of the nuclear installation and its safety features has been established.

"4.39. The direct and indirect pathways by which radioactive releases from the nuclear installation could potentially affect the public and the environment shall be identified and evaluated. In this evaluation, specific regional and site characteristics, including the population distribution in the region, shall be taken into account, with special attention paid to the transport and accumulation of radionuclides in the biosphere.

"4.40. It shall be demonstrated that the information provided to assess the potential effects on the population associated with accident conditions, including accidents that could warrant emergency response actions being taken in the external zone, is consistent with achieving the safety objective for site evaluation."

With regard to the dispersion of radioactive material, Requirement 25 of SSR-1 [2] states that **"The dispersion in air and water of radioactive material released from the nuclear installation in operational states and in accident conditions shall be assessed."**

Furthermore, associated paras 6.3–6.7 of SSR-1 [2] establish requirements that the dispersion of radioactive material in air and water during both operational states and accident conditions are to be assessed. This includes conducting hydrogeological and hydrological investigations to characterize the movement of radionuclides in surface water and groundwater, as well as their potential radiological impact. The assessment is supposed to be supported by a comprehensive survey and monitoring programme covering aspects such as surface water characteristics, groundwater interactions, radionuclide migration and retention, and associated exposure pathways. Full details of the specific requirements can be found in SSR-1 [2].

To appropriately comply with Requirement 12 of SSR-1 [2], it is important to consider Requirement 25 of SSR-1 [2], which relates specifically to the potential for the radioactive effluents to impact people and the environment.

The critical point to establish in the analysis of the movements in the hydrosphere is the interaction between surface waters and groundwater. In general, surface waters (e.g. lakes, seas, rivers, estuaries) are monitored, and data on these waters are relatively easy to collect. Monitoring programmes normally start as soon as a site is selected; therefore, by the time the PSAR is prepared, there are already several years of data available for this purpose.

The collection of groundwater data is more problematic. It is first necessary to understand the regional hydrogeology; a thorough literature survey on this topic is the first step. Near regional studies need to be performed at approximately 1:25 000 to 1:50 000 scale. These involve hydrogeological maps and cross sections from near regional geophysical studies as well as borehole information.

Hydrogeological information on a site vicinity scale (approximately 1:5000 to 1:10 000) needs to be collected from ad hoc geophysical studies and borehole information as part of the site evaluation investigations. Major aquifers in the site vicinity and their relationship with near regional and regional aquifers need to be identified. Information on tracer tests needs to be included in the database for this part of the review.

All this information is used to prepare a three dimensional hydrogeological model of the site vicinity. The reviewer needs to verify that the model is data driven and that different interpretations of data are considered in order to represent epistemic uncertainties.

In general, uncertainties related to geological information and, in particular, hydrogeological information are significant. Furthermore, parameters related to porosity and transmissibility are extremely variable. The flow of groundwater through the matrix of an aquifer and through cracks, fissures and fractures is very different, and this contributes to the parameter uncertainties.

The reviewer needs to verify that sufficient margins have been considered in dispersion and diffusion characteristics to address these uncertainties.

Requirement 20 (see Section 4.2.6 of this Safety Report) and associated paras 5.15–5.17 of SSR-1 [2] address the evaluation of flooding hazards and floods due to precipitation and other natural causes in particular. Paragraphs 5.15–5.17 of SSR-1 [2] state:

"5.15. The potential for flooding in the region surrounding the site due to one or more natural causes, such as storm surge, wind generated waves, meteorological tsunamis or seiches, or extreme precipitation — or due to a combination of such events that have a common cause or a relatively high frequency of occurrence — shall be evaluated.

"5.16. Appropriate meteorological, hydrological and hydraulic models shall be developed to derive the flooding hazards for the site, including secondary effects such as debris, ice and sediments. Where available, relevant information from studies of historic and prehistoric floods shall be used to inform estimates of the frequency and magnitude of riverine floods.

"5.17. The potential for instability of a coastal area or river channel due to erosion or sedimentation shall be investigated."

Consideration needs to be given to flood duration parameters such as forecasting time, warning time, duration of inundation, and flood recession time, as applicable, as these would potentially affect the implementation of emergency response actions.

Floods due to excessive precipitation can happen at any site, and meteorological data need to be collected to assess this hazard. These data need to cover as long a history as available. Data from similar neighbouring and climatically similar regions are used to expand the database using an ergodic approach. To evaluate the hazard of flash floods due to precipitation, it is important to collect data on extreme rates of precipitation, for which the available data may be less reliable than data related to average precipitation, which would contribute to uncertainties.

Modelling and extrapolation can use extreme value statistics. However, attention needs to be paid to constraining the results. In any model or extrapolation, climate change will be an important factor to consider. The reviewer needs to verify that sufficient margins have been considered in the models and extrapolations to include both aleatory and epistemic uncertainties.

Requirements regarding water waves induced by earthquakes or other geological phenomena are established in paras 5.18–5.20 of SSR-1 [2].

"5.18. The potential for tsunamis or seiches in the region that could affect the safety of the nuclear installation shall be evaluated. The potential for tsunamis or seiches from phenomena other than seismic sources (e.g. from submarine landslides) shall be evaluated, as appropriate for the region.

"5.19. The hazards associated with tsunamis or seiches shall be derived from historical records and any available information on prehistoric floods, as well as from physical and/or analytical modelling. Such hazards shall include potential draw-down and run-up [footnote omitted] that could result in physical effects on the site.

"5.20. The hazards associated with tsunamis or seiches shall be evaluated as appropriate for the region, using nearshore bathymetry and coastal topography, with account taken of any amplification due to the coastal configuration (including artificial structures)."

The database needed for the evaluation of a tsunami or seiche related hazard covers a very large area. Earthquakes that occur thousands of kilometres away from the site may generate significant tsunamis or seiches at the site. For example, the design basis tsunami height for the Fukushima Daiichi nuclear power plant was based on an earthquake that occurred in Chile in 1960 [21]. The reviewer needs to verify that the database includes all potential sources of tsunami and seiche in the sea, lake or ocean on which the site is located. In general, subduction zones are the major sources of tsunamis or seiches. However, some tsunamis are caused by non-tectonic dislocation of the sea floor, such as submarine landslides or turbidity currents. Phenomena such as submarine volcanoes and large coastal landslides (including glaciers) need to be checked as possible source of tsunamis. Seiches may be caused even by earthquakes that occur on land.

All pre-historical and historical data on coastal flooding need to be collected. The tsunami hazard analysis generally uses deterministic methods supported by probabilistic tsunami hazard analysis. For both these approaches, a far field and a near field propagation model are needed, based on bathymetric data. Near field bathymetric data need to have higher precision and are generally based on ad hoc marine investigations for the project.

The results of the tsunami hazard analysis need to be checked by paleo-tsunami analysis that generally extends 50–100 km on either side of the site.

Modelling is a very important and sensitive part of tsunami hazard analysis. Modelling of the tsunami source (i.e. amount of displacement, style of faulting, and volume of displaced material related to submarine landslides), modelling of the pathway of the tsunami waves (both far field and near field), and modelling of the recurrence process (if a probabilistic approach is used) are major components of the overall tsunami model and need to be carefully scrutinized by the reviewer. It also needs to be confirmed that the tsunami evaluation team has good coordination with the team performing the seismic hazard analysis.

Both the database and the modelling contain significant uncertainties. Estimating the size of tsunamigenic pre-historical and historical earthquakes is problematic because the damage information can be obtained only from the affected land area, which is a small fraction of the total region of deformation. Furthermore, it might not be easy to retrieve pre-historical and historical data from far away regions. There may also be significant uncertainties in the source and path modelling. For all of these reasons, the reviewer needs to check that these uncertainties are duly accounted for, considering also that coastal flooding is an event that may lead to a cliff edge effect in the safety analysis.

Requirements regarding floods and waves caused by the failure of water control structures can be found in paras 5.21–5.23 of SSR-1 [2], which state:

> "5.21. Upstream water control structures such as dams shall be analysed to determine the potential hazard associated with the failure of one or more of the upstream structures, including in combination with flooding from other causes.

"5.22. If a preliminary examination of the nuclear installation indicates that it would not be able to safely withstand the effects of the failure of one or more of the upstream water control structures, then the hazards associated with the nuclear installation shall be evaluated with the inclusion of such effects. Alternatively, such upstream structures shall be analysed by methods equivalent to those used in determining the hazards associated with the nuclear installation to demonstrate that the upstream structures could survive the event concerned.

"5.23. Flooding and associated phenomena caused by an accumulation of water due to a blockage of rivers upstream or downstream (e.g. caused by landslides or ice), or due to a change in land use, shall be considered."

The database necessary for assessing the flood hazard from the failure of upstream structures mainly consists of information used to evaluate the safety of these structures with respect to floods and earthquakes. The safety evaluation also involves the possibility of multiple dam failures. Information regarding the design of these structures is generally available from the owner or operator of the dam. If the dam or dams are old, then it is likely that their design does not conform to modern standards, and they may be likely to fail at much lower annual frequencies of exceedance than those related to the nuclear installation.

It is also important to understand the failure mode of the dam. Earth dams, concrete gravity dams and arch dams all have different modes of failure, and this may have a significant effect on the way in which the flood event develops and reaches the site of the nuclear installation.

If there is more than one dam upstream, the modelling and analysis needs to consider a variety of scenarios. For example, instantaneous failure of all of the dams might not generate the most unfavourable scenario.

The reviewer needs to check that assumptions regarding the characteristics of the dam(s) and the combination of failure scenarios represent a conservatism commensurate with the uncertainties related to the database and modelling.

4.5.7. Reviews for small modular reactors

4.5.7.1. *Ultimate heat sink*

UHS requirements for large nuclear power plants and SMRs may differ significantly, and therefore the data needed and the uncertainties involved in reviews for these different facilities may be very different. Some SMRs have passive means of extracting residual heat.

In general, it would be expected that a graded approach could be applied to data collection related to, and protection of, the UHS for SMRs (i.e. graded in comparison to large nuclear power plants). The extent of this grading will depend on the specific UHS design, which may be different for different SMRs.

4.5.7.2. *Flooding*

Flooding events cannot, in general, be screened out. Even if the design of the SMR is such that it is not located near a river, lake or sea, there is always a possibility of a flash flood, and the necessary data need to be collected and evaluated and a hazard analysis performed. Since flooding would lead to common cause failures involving more than one module, ample margins are needed in the analysis.

Marine based SMRs are particularly vulnerable to coastal flooding, and the reviewer needs to verify that it is possible to implement engineering measures against such phenomena as tsunamis, seiches and storm surges. Large protection structures, such as sea walls, might not be cost-effective. In this case, the coastal flooding phenomena may potentially be exclusionary.

4.5.7.3. *Transport and diffusion of effluents in the hydrosphere*

Some SMRs may have very different characteristics from large nuclear power plants with regard to any liquid waste that is routinely discharged or accidentally released. For this reason, it may be appropriate to apply a graded approach to the investigations into the hydrology and hydrogeology of the region. For marine based SMRs, hydrogeological investigations would not be necessary.

The reviewer needs to check all possible scenarios involving the release of liquid effluents. It is also important to have an understanding of the radionuclides involved. The pathways by which these effluents may reach surface water and groundwater then need to be reviewed.

4.5.8. References for the review

The references for Section 4.5 include SSG-61 [1], SSR-1 [2], NS-G-3.2 [9], SSG-35 [10] and SSG-18 [13].

4.6. METEOROLOGY

4.6.1. Areas of review

The target of the SAR review in Section 4.6 includes the information recommended in paras 3.2.25–3.2.27 of SSG-61 [1].

There are three distinct areas of review for meteorology. The first two are related to hazards associated either with an extreme value of a meteorological parameter or with the occurrence of rare hazardous phenomena, as follows:

(a) Extreme meteorological parameters include:
   (i)   Air temperature;
   (ii)   Wind speed;
   (iii) Precipitation (liquid equivalent);
   (iv)   Snowpack.
(b) Rare meteorological phenomena include:
   (i)   Lightning;
   (ii)   Tropical cyclones, typhoons and hurricanes;
   (iii) Tornadoes;
   (iv)   Waterspouts.

These phenomena are considered external hazards and relate to the impact of the environment on the nuclear installation.

For the evaluation of hazards due to these events, the first important step is the preparation of an appropriate database. Recommendations related to the database are given in Section 3 of SSG-18 [13]. The simultaneous occurrence of some of these events may also be possible and needs to be considered. In addition, the effects of climate change need to be considered; this could affect both extreme meteorological parameters and rare meteorological phenomena, pushing extreme values to new levels. For rare meteorological events, both their frequency and scale may increase. Moreover, rare phenomena may start occurring where they were not experienced before.

The third area of review is the transport and diffusion of effluents discharged into the atmosphere. Section 2 of NS-G-3.2 [9] provides recommendations on the database and methods necessary to evaluate the transport and diffusion of effluents into the atmosphere.

Requirements 6, 7, 10, 12, 18, 19 and 25 of SSR-1 [2], and the recommendations provided in SSG-18 [13] and NS-G-3.2 [9], are the most relevant for these areas of the review.

Section 4.6 interfaces with Sections 4.2, 4.5, 4.8, 4.10 and 4.11 of this Safety Report.

The paragraphs in SSG-61 [1] relevant to these areas of the review state:

"3.2.25. This Section should provide a description of the meteorological aspects relevant to the site and its surrounding area, with account taken of regional and local climatic effects. Data derived from on-site meteorological monitoring or other meteorological stations should be documented.

"3.2.26. This Section should include information relevant to the assessment of (i) the hazards from meteorological events potentially affecting the plant and (ii) the transport of radioactive material to and from the site and the dispersion of radionuclides through the environment.

"3.2.27. The extreme values of meteorological parameters or meteorological events — such as temperature; humidity; rainfall; wind speeds for straight and rotational winds, including tornadoes (owing to the sudden pressure drop that accompanies the passage of the centre of a tornado); waterspouts (owing to their potential to transfer large amounts of water to the land from nearby water bodies); dust storms; sandstorms; and snow loads and ice (see SSG-18 [13]) — should be evaluated in relation to the design, with account taken of the envisaged evolution of such extreme parameters over the lifetime of the nuclear power plant. The potential for lightning and windborne debris to affect plant safety (including the design basis missile hazard from hurricanes and tornadoes) should be considered, where appropriate."

4.6.2. Acceptance criteria

Extreme values of meteorological parameters are based on meteorological measurements. Acceptance criteria for the annual frequency of exceedance corresponding to the design basis for these parameters need to be provided in national regulations. Generally, local meteorological observations are too short to extrapolate to very low exceedance frequencies, and the acceptance criteria can be met only by using data imported from nearby regions with similar characteristics. Similarly, for rare meteorological phenomena, the annual frequency of exceedance corresponding to the design basis needs to be provided in national regulations. If these acceptance criteria are not given in national regulations, they need to be discussed and agreed with the regulatory body.

Discharges during normal operation and during accident conditions need to be considered. Acceptance criteria (for both normal operation and accidental discharges) are generally not related to transport or diffusion properties but to the concentration of the transported radionuclides, which is addressed in Section 4.8 of this Safety Report.

4.6.3. Review procedure

A typical safety review of an SAR consists of a desktop review of documents submitted by the applicant, site visits to confirm consistency between the SAR and the site condition, and independent calculations for verification.

The review focuses both on the soundness of the database and on the approaches used in the hazard analysis. Both analytical and physical models may be used, and therefore the review needs to also consider the way in which epistemic uncertainties are treated.

A site visit is necessary in order to understand the nuclear installation layout and the locations of the instruments of meteorological measurement and their elevations.

Independent checks of some hazard calculations may also be needed particularly when the results, even after the necessary sensitivity and/or uncertainly analysis, indicate that acceptance criteria are met with only small margins.

For the transport and diffusion of effluents, the review will involve detailed models of atmospheric dispersion. Mathematical models need to be scaled to real measurements, and therefore the database for these measurements needs to be reviewed. A site visit is necessary in order to understand the potential points of discharge and the planned points of measurements. The validation and verification of the computer programs used need to also be reviewed.

4.6.4. Evaluation findings

The reviewer's evaluation of site characteristics, in accordance with the relevant regulatory criteria, needs to be documented in the SER, noting any findings that diverge from these criteria. The first step of the review is a decision on the overall quality and completeness of this Section of the SAR. It then needs to be verified that there are no issues related to site suitability.

The documentation of the review needs to include the evaluation of meteorological hazards with respect to the relevant regulatory criteria. The reviewer is responsible for independently verifying that the contents of the SAR comply with the acceptance criteria and for ensuring that the conclusions of the evaluation report also meet the acceptance criteria. The procedure followed in the evaluation needs to be briefly described. If needed, the evaluation may include the performance of independent calculations to ascertain compliance.

The SER needs to include short descriptions of the approaches adopted for meteorological hazard analyses, including the database used and the results obtained. This includes the target probabilities used in the design basis and beyond design basis for meteorological hazards.

It is necessary to verify that the applicant has provided sufficient information and that the review and calculations (if applicable) support the conclusions of the SER.

With regard to the transport and diffusion of effluents in the atmosphere, compliance with acceptance criteria is discussed in Section 4.8 of this Safety Report.

4.6.5. Implementation

The Requirements established in SSR-1 [2] and the recommendations provided in SSG-18 [13] and NS-G-3.2 [9] need to be used in the evaluations performed by the applicant, together with all applicable national regulations. If national safety targets and acceptance criteria related to meteorology are complied with using different approaches from those provided in the national regulations and IAEA safety standards, the methods proposed by the applicant need to be duly scrutinized.

4.6.6. Detailed review of technical content

Meteorology is associated with several potential impacts on the safety of a nuclear installation: the dispersion of radioactive effluents in the atmosphere, extreme meteorological hazards and rare meteorological phenomena.

Requirements 12, 18, 19 and 25 of SSR-1 [2] address these topics, and they will be treated separately in this Safety Report and need to be reviewed accordingly.

The requirements regarding the potential effects of the nuclear installation on people and the environment, established in Requirement 12 and associated paras 4.38–4.40 of SSR-1 [2], are given in Section 4.5.6 of this Safety Report.

The requirements regarding the dispersion of radioactive material, and the atmospheric dispersion of radioactive material in particular, are established in Requirement 25 and associated paras 6.1 and 6.2 of SSR-1 [2], which state:

"The dispersion in air and water of radioactive material released from the nuclear installation in operational states and in accident conditions shall be assessed.

"Atmospheric dispersion of radioactive material

"6.1. The analysis of the atmospheric dispersion of radioactive material shall take into account the orography, land cover and meteorological features of the region, including parameters such as wind speed and direction, air temperature, precipitation, humidity, atmospheric stability parameters, prolonged atmospheric inversions and any other parameters required for modelling of atmospheric dispersion. If possible, long term meteorological data for nearby locations shall be obtained, evaluated for quality and used.

"6.2. A programme for meteorological measurements shall be prepared and carried out at or near the site using instrumentation capable of measuring and recording the main meteorological parameters at appropriate elevations, locations and sampling intervals. Data from at least one representative full year shall be collected and used in the analyses of atmospheric dispersion, together with any other relevant data available from other information sources. The meteorological data shall be expressed in terms of appropriate meteorological parameters."

The data for the purposes of these two requirements are collected generally from the meteorological tower, placed at or very near the site, with a height at least equal to the highest release point of the nuclear installation. The database needs to cover at least one full year and all seasonal variations.

Normal and extreme values of climate parameters — such as air pressure, air temperature (or dry bulb temperature and humidity), precipitation, and wind speed and direction — characterize the meteorological environment. The data to be collected include the wet bulb temperature (which can be calculated as a function of the dry bulb temperature), the dew point temperature (or relative humidity) and the air pressure. Wind data are generally presented in the form of wind roses superimposed on the demographic data of the near region (see also Section 4.1 of this Safety Report).

Dispersion models are based on the local meteorological data obtained from the meteorological tower, and they need to cover a distance in excess of the external zone of the nuclear installation.

Uncertainties may be related to instrument errors, and/or data processing, which can be minimized through an effective QA programme. The reviewer needs to verify that uncertainties associated with instrument errors or data processing have not adversely affected the data. Another source of uncertainty is related to the modelling. The reviewer needs to ascertain that these uncertainties are accounted for and that the conservatism of the analysis is sufficient to consider them.

Requirement 18 (see Section 4.2.6 of this Safety Report) and associated paras 5.11 and 5.12 of SSR-1 [2] address the evaluation of extreme meteorological hazards. Paragraphs 5.11 and 5.12 of SSR-1 [2] state:

"5.11. Meteorological phenomena such as wind, precipitation, snow and ice, air and water temperature, humidity, storm surges and sand or dust storms, as well as their credible combinations, shall be evaluated for their extreme values [footnote omitted] based on available records. If necessary, efforts shall be made to extend the database on meteorological hazards (e.g. by incorporating historical climate data, numerical models and simulations).

"5.12. Appropriate methods shall be applied for the evaluation of meteorological hazards, taking into account the amount of data available (both measured data and historical data) and known past changes in relevant characteristics of the region."

All data related to extreme meteorological hazards need to be accompanied by explanatory information on the data (metadata). The reviewer needs to verify that such metadata are provided and are adequate. In the processing of the data, the non-stationary behaviour of the stochastic process under consideration, which may reflect climatic variability and climate change, among other phenomena, needs to be considered. Paragraph 3.16 of SSG-18 [13] states "Most national meteorological services publish catalogues listing the specific meteorological and climatic data they have collected, including data on wind, temperature and precipitation."

Some variations in field measurements may introduce uncertainties in the database, which need to be taken into consideration. Regarding these variations, para.3.16 of SSG-18 [13] states:

"Users of these data should be aware that while national meteorological services generally follow standards for measurement that are established by the WMO, field measurements made by different organizations for meeting different requirements do not necessarily follow the same standards. For instance:

(a)   The standard 10 m height and instrument exposure for measuring wind speed and direction might not be implemented owing to the logistics of instrument installation.

(b)   Measurement techniques for recording maximum wind speed vary from State to State. The general tendency is to record average values for a given constant duration, such as 3 s gusts, 60 s averages or 10 min averages (the averaging time is a characteristic of the database).

(c)   Air temperatures (such as dry bulb and dew point temperatures) are recorded continuously at some recording stations and at frequent intervals at other stations. At some secondary locations, only the daily maximum and minimum air temperatures are recorded.

(d) Data that are routinely collected and used for analyses of extreme maximum precipitation generally include the maximum 24 h precipitation depth. Records based on shorter averaging times contain more information and should under certain circumstances be preferred [footnote omitted].

This variation necessitates careful evaluation and, if possible, adjustment of the data before processing. Such information, including information on the data processing methods used, should be documented."

Guidance on the reporting of meteorological data and the application of numerical models in the evaluation of meteorological site characteristics is provided in IAEA Safety Standards Series publication No. SSG-18 [13].
Paragraph 3.17 of SSG-18 [13] states:

"A report on the results of the analyses should include a description of each meteorological station and the monitoring programme, including: types of instrument, calibration history, geographical location, instrument exposure and altitude, data record period(s) and data quality."

Lack of data from the region of the site may be compensated by the use of data from neighbouring similar regions using the ergodic assumption. However, doing so may introduce additional uncertainties to the database.
Paragraph 3.18 of SSG-18 [13] states:

"Numerical mesoscale models with spatial resolution adequate to resolve the regional and local geophysical features of the site are useful for simulating the atmospheric circulation and other local meteorological parameters at regional and local scales."

Extreme annual values of meteorological parameters constitute examples of random variables, which may be characterized by specific probability distributions. In principle, the data set is analysed with appropriate probability distribution functions. Among these functions, the following generalized extreme value distributions are widely used: Fisher-Tippett Type I (Gumbel), Type II (Fréchet) and Type III (Weibull).
The reviewer needs to verify that the database considers all of the available sources and includes all uncertainties in the model. The probability functions employed need to be data driven and also sufficiently constrained so that reasonable extrapolations are possible. As the meteorological processes will

tend to be non-stationary due to climate change, the reviewer needs to check that these effects have been appropriately dealt with and that the results are sufficiently conservative.

Requirement 19 (see Section 4.2.6 of this Safety Report) and associated paras 5.13 and 5.14 of SSR-1 [2] address the evaluation of extreme meteorological hazards. Paragraphs 5.13 and 5.14 of SSR-1 [2] state:

"Lightning

"5.13. The potential for the occurrence and the frequency and severity of lightning shall be evaluated for the site vicinity.

"Tornadoes and cyclones

"5.14. The potential for the occurrence and the frequency and severity of tornadoes, cyclones and associated missiles shall be evaluated for the site. The hazards associated with tornadoes and cyclones shall be derived and expressed in terms of parameters such as rotational wind speed, translational wind speed, radius of maximum rotational wind speed, pressure differentials and rate of change of pressure."

For rare meteorological phenomena, including lightning, tropical cyclones, typhoons, hurricanes, tornadoes and waterspouts, the specific meteorological data on storm parameters need to be collected. Paragraph 4.45 of SSG-18 [13] states:

"The following data on the storm parameters for tropical cyclones should be collected:

— Minimum central pressure;
— Maximum wind speed;
— Horizontal surface wind profile;
— Shape and size of the eye;
— Vertical temperature and humidity profiles within the eye;
— Characteristics of the tropopause over the eye;
— Positions of the storm at regular, preferably six-hourly, intervals;
— Sea surface temperature."

It is also important that the normal or 'undisturbed' conditions prevailing in the region are known. Paragraph 4.47 of SSG-18 [13] presents the climatological charts or analytical results to be reviewed for this purpose, as follows:

"— Sea level pressure;
— Sea surface temperature;
— Air temperature, height and moisture (dew points) at standard pressure levels and at the tropopause."

Rare meteorological events are unlikely to be recorded at any single location or by a standard instrumentation network owing to their low frequency of occurrence. In addition, such events could damage standard instruments or cause unreliable measurements. For rarely occurring phenomena, such as phenomena that produce extreme wind speeds, an estimation of the intensity of the phenomenon needs to be determined on the basis of conceptual or numerical models of the phenomenon, coupled with statistical methods appropriate for the rate of occurrence and the intensity of the event at the nuclear installation site. The size of the region to be investigated depends on the specific characteristics of the meteorological and geographical environment of the area in which the site is located and the hazard under consideration (e.g. tornadoes, hurricanes).

Two types of data, which are generally available from national meteorological services, are collected for rare meteorological phenomena:

(a) Data and information systematically collected, processed and analysed in recent years, which may include more occurrences of events of lower intensity and may be more reliable than historical (anecdotal) data;
(b) Historical data.

Occasionally, a comprehensive collection of data and information obtained soon after the occurrence of a rare meteorological event may be available. This could include measured values of variables, eyewitness accounts, photographs, descriptions of damage and other qualitative information that were available shortly after the event. Such detailed studies of actual events are important in constructing a model for their occurrence.

Following the collection of data on rare meteorological phenomena, a specific dedicated catalogue needs to be compiled with an appropriate check for completeness.

The general procedure for evaluating the hazard associated with the occurrence of rare meteorological events comprises the following:

(a) A study of the representative data series available for the region under analysis and an evaluation of its quality (representativeness, completeness, effectiveness of the QA programme, and homogeneity);
(b) Selection of the most appropriate statistical distribution(s) for the data set;
(c) Processing of the data to evaluate the moments of the probability distribution function of the parameter under consideration (expected value, standard deviation and others, if necessary), from which the mean recurrence interval and associated confidence limits may be estimated.

Both the database and the modelling of rare meteorological phenomena involve significant uncertainties. A recently acquired database has a short duration and therefore is not likely to include maximum credible events, whereas the historical data are not likely to be as reliable as recent data.

Modelling through the use of statistical distributions necessarily makes stationarity assumptions, which is most likely to be biased to the unconservative side.

The reviewer needs to check the soundness of the database as well as the uncertainties associated with the database and the modelling to verify that sufficient margins have been incorporated to take these into account.

4.6.7. Reviews for small modular reactors

4.6.7.1. Potential effects of the nuclear installation on people and the environment and dispersion of radioactive material

The dispersion of radionuclides in the atmosphere and the effects on the people and environment may be different for SMRs from those for large nuclear power plants.

All possible scenarios for the SMR to release or discharge radioactive effluents into the atmosphere need to be reviewed, together with the points of release. It is very likely that releases to the atmosphere would be from lower heights than they would at a large nuclear power plant due to the difference in size. This would mean that the data requirements may be significantly different and the "standard 10 m mast" [13] might not be necessary at the SMR site. The review of the atmospheric dispersion needs to be conducted using performance based principles.

Once again, the great variability among the different SMRs needs to be underlined.

4.6.7.2. Evaluation of extreme meteorological hazards

In principle, there would not be any difference in the reviews of a large nuclear power plant and an SMR, and the discussion in Section 4.6.6 of this Safety Report also applies to SMRs. The sensitivity of a large nuclear power plant to air temperature depends on its design; therefore, this aspect may need special attention in the review process.

4.6.7.3. Evaluation of rare meteorological events

In principle, there would not be any difference in the review of a large nuclear power plant and an SMR, and the discussion in Section 4.6.6 also applies to SMRs.

4.6.8. References for the review

The references for Section 4.6 include SSG-61 [1], SSR-1 [2], NS-G-3.2 [9], SSG-35 [10] and SSG-18 [13].

4.7. GEOLOGY, SEISMOLOGY AND GEOTECHNICAL ENGINEERING

4.7.1. Areas of review

The target of the SAR review in Section 4.7 includes the information recommended in paras 3.2.28–3.2.31 of SSG-61 [1].

There are four distinct topics in this review area:

(a) Vibratory ground motion hazard;
(b) Fault displacement hazard;
(c) Geotechnical hazards, including liquefaction, collapse, subsidence and slope instability (including those induced by seismic ground motion);
(d) Volcanic hazards.

Recommendations on vibratory ground motion hazards and fault displacement hazards are provided in SSG-9 (Rev. 1) [12]; recommendations on geotechnical hazards are provided in NS-G-3.6 [19]. Volcanic hazards are not addressed in SSG-61 [1]; however, recommendations on the evaluation of volcanic hazards are provided in SSG-21 [11].

The areas of review consist of the appropriate compilation of the database, modelling, accounting for aleatory and epistemic uncertainties, and hazard calculation.

Requirements 4, 6, 7, 15–17, 21 and 22 of SSR-1 [2] are the most relevant for these areas of the review.

Section 4.7 interfaces with Sections 4.2 and 4.11 of this Safety Report.

The paragraphs in SSG-61 [1] relevant to these areas of review state:

"3.2.28. This Section should provide information concerning the geological, tectonic, seismological and volcanic characteristics of the site and a sufficiently large region surrounding the site. The evaluation of seismic hazards should be based on a suitable seismotectonic model, substantiated by appropriate seismological evidence and geological or seismological data. The results of this evaluation that will be used further in other sections of the safety analysis report (including structural design and seismic qualification of components) should be described in sufficient detail. The potential for volcanic phenomena to affect plant safety should be considered, where appropriate.

"3.2.29. Site reference data relating to the geotechnical properties of soil and rock underlying the site (both static and dynamic properties including damping and modulus degradation properties) should be elaborated on in this section. Geological hazards such as slope instability, subsidence or uplift of the site surface, soil liquefaction, instability of subsurface materials, and the long term performance of subsurface materials and foundations over the lifetime of the plant should be characterized in this section. The processes for the following should be described: the collection of data for the design of foundations, the evaluation of the effects of site response and soil–structure interaction, the construction of earth structures and buried structures, the evaluation of the effects of groundwater conditions, and the evaluation of soil improvements at the site.

"3.2.30. This Section should present the relevant data for the site and the associated ranges of uncertainty, including the spatial variability used in the site seismic response analysis and in the structural design. Reference should be made to the technical reports that provide a detailed description of the conduct of the investigations and their planned extensions, as well as of the origin of the data collected through site surveys on a regional basis or through bibliographic surveys.

"3.2.31. The design of subsurface material and of buried structures, as well as site protection measures, if relevant, should also be documented. A description of projected developments relating to the information described in paras 3.2.28–3.2.30 should also be provided and should be updated as required."

4.7.2. Acceptance criteria

National regulations may include more quantitative acceptance criteria (e.g. a limiting bearing capacity for foundation soils), and these need to be complied with.

The IAEA safety standards specify qualitative criteria for assessing site acceptability, taking into account the site–design combination. Quantitative criteria in relation to fault capability and volcano capability are recommended in SSG-9 (Rev. 1) [12] and SSG-21 [11], respectively. Fault capability with associated criteria, is defined in Section 8 of SSG-9 (Rev. 1) [12]; volcano capability is defined in para. 2.19 and fig. 1 of SSG-21 [11].

The criteria for fault capability are dependent on the time frame chosen, which in turn depends on the seismotectonic setting of the site.

Massive liquefaction may be an exclusionary hazard; however, it is always necessary to check the availability of engineering solutions for liquefaction before deciding on site acceptability.

4.7.3. Review procedure

A typical safety review of an SAR consists of a desktop review of documents submitted by the applicant, site visits to confirm consistency between the SAR and the site condition, and independent calculations for verification.

The review of vibratory ground motion hazards mainly involves desktop reviews, including review of documents and checking of calculations. As modelling is a key activity in this analysis, a review of the way in which epistemic uncertainties are treated is essential.

The reviews of fault capability and fault displacement hazards definitely involve field visits (which may involve trench inspections) as well as a review of laboratory testing techniques if radiometric age dating has been used.

Geotechnical hazards also require multifaceted reviews based on documentation, field visits and laboratory analyses. There is significant interface between the vibratory ground motion hazards and geotechnical engineering and geotechnical hazards, and these interfaces need special attention from the reviewer. Site soil conditions affect the vibratory ground motion hazards, and the level of the ground motion affects the onset of geotechnical hazards such as liquefaction and slope instability.

If only the distant effects of a volcano are present for the site, a review of volcano induced hazards based on documentation would be sufficient. However, if the volcano is relatively close to the site so that some of the exclusionary effects need to be checked (see table 1 of SSG-21 [11]), a site visit is necessary. A review of laboratory techniques may also be needed if radiometric age dating has been used.

If calculations show, even after the necessary sensitivity and/or uncertainly analysis, that acceptance criteria are too close to the findings (i.e. there are small margins), then independent calculation checks would be needed.

4.7.4. Evaluation findings

The reviewer's evaluation of site characteristics, in accordance with the relevant regulatory criteria, needs to be documented in the SER, noting any findings that diverge from these criteria. The first step of the review is a decision on the overall quality and completeness of this Section of the SAR. It then needs to be verified that no site suitability issues exist, using the acceptance criteria discussed in Section 4.7.2 of this Safety Report.

The documentation of the review needs to include the evaluation of vibratory ground motion and fault displacement hazards with respect to the relevant regulatory criteria. The reviewer is responsible for independently verifying that the contents of the SAR comply with the acceptance criteria and for ensuring that the conclusions of the evaluation report also meet the acceptance criteria. The procedure followed in the evaluation needs to be briefly described. If needed, the evaluation may include independent calculations to ascertain compliance.

The documentation of the review needs to include the evaluation of geotechnical hazards and their relationship with vibratory ground motion hazards. Volcanic hazard evaluation also needs to be included (if not screened out).

The SER needs to include short descriptions of the approaches adopted for the hazard analyses, including the database used and the results obtained. This includes the target probabilities used in the design bases and beyond design bases for these hazards.

It is necessary to verify that the applicant has provided sufficient information and that the review and calculations (if applicable) support the conclusions of the SER.

4.7.5. Implementation

The requirements established in SSR-1 [2] and the recommendations provided in SSG-9 (Rev. 1) [12], SSG-21 [11], SSG-61 [1] and NS-G-3.6 [19] need to be used in the evaluations performed by the applicant, together with all applicable national regulations. If national safety targets and acceptance criteria related to geology, seismology and geotechnical engineering are complied with using different approaches from those set forth in the national regulations and IAEA safety standards, the methods proposed by the applicant need to be duly scrutinized.

4.7.6. Detailed review of technical content

The topic of geology, seismology and geotechnical engineering is very wide and is associated with several potential important impacts on the safety of a nuclear installation. Requirement 4 of SSR-1 [2] states that **"The suitability of the site shall be assessed at an early stage of the site evaluation and shall be confirmed for the lifetime of the planned nuclear installation."** Associated requirements (i.e. paras 4.6–4.9 of SSR-1 [2]) are presented in Section 3.1 of this Safety Report.

Requirement 7 (see Section 3.1 of this Safety Report) and associated paras 4.20 and 4.21 of SSR-1 [2] address external hazard evaluation. Paragraphs 4.20 and 4.21 of SSR-1 [2] state:

"4.20. The site evaluation for a nuclear installation shall consider the frequency and severity of natural and human induced external events, and potential combinations of such events, that could affect the safety of the nuclear installation.

"4.21. Information on the frequency and severity of external events derived from the characterization of the hazards shall be used in establishing the site specific design parameters. Adequate account shall be taken of both aleatory uncertainties and epistemic uncertainties in the establishment of site-specific design parameters."

The geological and geotechnical conditions of a nuclear installation site may have a significant impact on site acceptability and need to be assessed early in the process. The requirements related to site acceptability cover fault capability, volcanic hazards and geotechnical hazards, and they are provided under Requirements 15 (Evaluation of fault capability), 17 (Evaluation of volcanic hazards) and 22 (Evaluation of geotechnical hazards and geological hazards) of SSR-1 [2].

Requirement 15 (see Section 4.2.6 of this Safety Report) and associated paras 5.2–5.4 of SSR-1 [2] address external hazard evaluation. Paragraphs 5.2–5.4 of SSR-1 [2] state:

"5.2. Capable faults [footnote omitted] shall be identified and evaluated. The evaluation shall consider the fault characteristics in the site vicinity. The methods used and the investigations made shall be sufficiently detailed to support safety related decisions.

"5.3. The potential effect of fault displacement on safety related structures, systems and components shall be evaluated. The evaluation of fault displacement hazards shall include detailed geological mapping of excavations for safety related engineered structures to enable the evaluation of fault capability for the site.

"5.4. A proposed new site shall be considered unsuitable when reliable evidence shows the existence of a capable fault that has the potential to affect the safety of the nuclear installation and which cannot be compensated for by means of a combination of measures for site protection and design features of the nuclear installation. If a capable fault is identified in the site vicinity of an existing nuclear installation, the site shall be deemed unsuitable if the safety of the nuclear installation cannot be demonstrated."

As the issue of external hazard evaluation is related to site acceptability, it needs to be addressed early in the site survey and site selection process. However, the very detailed site investigations that may be needed for fault capability assessment (see Section 4.7.6.2 of this Safety Report) are not feasible at this stage of the site survey.

4.7.6.1. Fault capability assessment during site survey and site selection process

The database needed for the site survey and site selection stage need to include the following:

(a) Geological and geomorphological maps and cross sections of the near region at a scale of 1:25 000 to 1:50 000;
(b) Map of capable faults (if available) or tectonic map of the near region at a scale of 1:25 000 to 1:50 000;

(c)   Seismological data from local networks (if available);
(d)   Geophysical profiles in the near region (if available), which could be gravity, magnetic, electrical or seismic profiles;
(e)   Any information on age dating of the formations in the near region;
(f)   Remote sensing information, such as aerial photographs, lidar data and satellite imagery.

Depending on the size of the region of interest for the study, the data from this list may need to be collected at a near regional scale.

The analysis involves mapping all of the potential capable faults on a single map and identifying the areas to be excluded on the basis of the distance from these faults. The exclusion of capable faults in the site vicinity would be a reasonable criterion for this step (see paras 7.10 and 7.11 of SSG-9 (Rev. 1) [12]). However, given that faults are three dimensional and may dip towards the site, as well as the other uncertainties involved, a 10 km exclusion zone may be appropriate.

If uncertainties in the database are large and one or more of the items listed earlier in this Section of the Safety Report is not available, it is possible that many non-capable faults could be designated as 'capable' in this step, and the available suitable area may decrease in size significantly. Although this approach can be considered conservative from a geological point of view, it is also possible that otherwise suitable sites would be eliminated through this process. Therefore, a balanced approach is needed at this step.

The uncertainties involve the following:

(a)   Identification and location of faults;
(b)   Age dating;
(c)   Appropriate selection of the fault capability time frame.

4.7.6.2.   Fault capability assessment during site evaluation

As well as being conducted prior to construction, site evaluation may be repeated during the operation of a nuclear installation when a suspect feature that has the potential to affect the safety of the installation is discovered in the site vicinity.

Recommendations on the investigations necessary to determine the fault capability are clearly indicated in paras 7.6–7.9 of SSG-9 (Rev. 1) [12], as follows:

"7.6. Sufficient surface and subsurface related data should be obtained from the investigations in the regional, near regional, site vicinity and site areas (see

Section 3) to demonstrate the absence of faulting at or near the site or, if faults are present, to describe the direction, extent, history and rate of movement of these faults as well as the age of the most recent movement.

"7.7. When surface faulting is known or suspected to be present, investigations should be conducted at the site vicinity scale and should include very detailed geological and geomorphological mapping, topographical analyses, geophysical surveys (including geodetic measurements, if necessary), trenching, boreholes, age dating of sediments or faulted rock, local seismological investigations, and any other appropriate and state of the art techniques (e.g. remote sensing methods) to ascertain the amount and age of previous displacements or deformations.

"7.8. Consideration should be given to the possibility that faults that have not shown recent near surface movement might be reactivated by human activity (e.g. reservoir loading, fluid injection, fluid withdrawal).

"7.9. Investigations of a capable fault should be sufficient to enable a confident decision to be made regarding whether the fault can be screened out as a credible hazard to nuclear safety or, if the hazard is judged to be credible, to provide sufficient quantitative information to the subsequent site evaluation, design and safety analysis process in accordance with para. 10.3. The capable fault investigations should also link to those investigations undertaken for vibratory ground motion analysis and should be consistent with them. While the specific needs of both analyses are somewhat inconsistent in terms of data needs and outputs, the documented narrative that reports on these analyses should recognize that both hazards derive from the same tectonic structures in the region."

A potential capable fault within or close to the site vicinity may require a major effort to resolve. Therefore, a serious attempt to make sure that no such fault exists within or close to the site vicinity is worthwhile.

Depending on the tectonic setting and the geological conditions in the site vicinity, it is possible to optimize the selection of methods; the reviewer needs to take this into consideration. For example, if the applicable time frame for capability is late Pleistocene (~125 000 years), and the pre-Quaternary bedrock is only a few metres deep, then the evidence for capability is constrained to the first few metres, in which case remote sensing data coupled with trenching would be the most suitable methods. However, for the same applicable time frame for capability, if the pre-Quaternary bedrock is located at 100 m or deeper and covered with Pleistocene and Holocene sediments, then geophysical methods would be more effective tools for resolving the issue.

Section 7 of SSG-9 (Rev. 1) [12] needs to be well understood and implemented, as there are different criteria for new versus existing nuclear installations. Also, a clear distinction needs to be made between primary (seismogenic) faults and secondary or distributed faulting that may manifest surface expression but is not seismogenic. These concepts are very important, especially when a capable fault is found to exist in the site vicinity and a fault displacement hazard analysis is necessary to demonstrate the safety of the nuclear installation.

If a fault displacement hazard analysis has been performed, the reviewer needs to check that the capable fault in question is a secondary (non-seismogenic) fault. The data needed for a fault displacement hazard analysis are the recurrence characteristics of the capable fault, comprising well constrained data on age dating and paleo-seismology using all available tools. It is important to reduce uncertainties to a minimum. One way to achieve this is to make sufficient data available so that all relevant experts involved in the project reach consensus regarding the issue. The reviewer needs to check that the epistemic uncertainties have been minimized.

Requirement 16 of SSR-1 [2] addresses the evaluation of ground motion hazards (see Section 4.2.6 of this Safety Report). The associated requirements established in para. 5.5 of SSR-1 [2] highlight the combination of vibratory ground motion effects and seismicity from human activities, such as dam construction, mining, and oil or gas well operations.

Two approaches for the evaluation of vibratory ground motion hazard are recommended in Section 6 of SSG-9 (Rev. 1) [12]: probabilistic and deterministic. Probabilistic seismic hazard analysis is generally the preferred method and is always needed as input to the seismic probabilistic safety assessment. The database needed for the two approaches is similar. SSG-9 (Rev. 1) [12] provides a very structured framework for the geological, geophysical and geotechnical database, which follows spatial considerations, as well as for the seismological database, which is temporally arranged.

The four scales of study for the geological, geophysical and geotechnical database are: regional (radius of several hundred kilometres), near regional (~25 km radius), site vicinity (minimum 5 km radius) and site area (inside the boundary of the nuclear installation). Recommendations on the use of these areas in the hazard evaluation are provided in paras 3.16–3.35 of SSG-9 (Rev. 1) [12]. Paragraph 3.17 of SSG-9 (Rev. 1) [12] states:

"Thus, the extent of the geographical area of interest at a regional scale should be defined in accordance with the recommendations provided in para. 3.6 and by considering the potential sources of all hazards generated by earthquakes that might affect the safety of the nuclear installations at the selected site. The size of the region to be investigated when assessing vibratory ground motion hazards should be large enough to incorporate all seismogenic structures that could affect the nuclear installation: the extent of this region is typically a few hundred kilometres in radius".

The reviewer needs to check that all credible seismic sources that may have an impact on the ground motion at the site have been considered for all structural frequencies and for the levels of annual exceedance probabilities that are used in the probabilistic seismic hazard analysis. Particular attention needs to be paid to subduction zones, which may be ~1000 km from the site but may still have an impact on the seismic hazard. However, unless the regulatory body has specific requirements for the radius to be considered, it might not be useful to model unnecessary sources as they may unduly burden the logic tree.

In general, regional scale studies are desktop studies and would not necessarily involve field or laboratory investigations. However, there may be features in the region that need particular attention, and such investigations are generally based on near regional studies. The same principle holds for the near region and the site vicinity, where more detailed investigations may be necessary for some features than would normally be needed.

Paragraphs 3.36–3.59 of SSG-9 (Rev. 1) [12] provide recommendations on the seismological database. Seismological data are expected to be collected and presented in temporal scales: paleo-seismological, archaeo-seismological, historical, instrumental and local instrumental. Data from older periods are generally used to constrain the maximum magnitudes (and sometimes their frequencies) associated with seismic sources; data from newer periods are used in calculating recurrence rates.

The reviewer needs to verify that a project specific catalogue has been compiled using all of the information in the seismological database. The processing of the seismological data is a very important step before those data can be used in the hazard analysis. Aleatory uncertainties in the database need to be accounted for.

The modelling needed for a probabilistic seismic hazard assessment (PSHA) or a deterministic seismic hazard assessment (DSHA) needs to take into account epistemic uncertainties; consequently, it may involve multiple groups of experts to represent the centre, body and range of technically defensible interpretations. The reviewer needs to be familiar with the concept and check that both the seismic source modelling and the ground motion modelling has satisfied this requirement. This attribute of a PSHA and a DSHA makes the study defensible and transparent and generally ensures that the results will be valid for a significant period of time.

The logic tree approach is generally used to represent the epistemic uncertainties. This is valid both for the PSHA and the DSHA. The software used in the PSHA needs to be checked to verify that it has proper documentation for verification and validation. The reviewer also needs to be familiar with the expected output of the PSHA and/or DSHA, which needs to be explicitly stated in the study. Section 10 and the annex of SSG-9 (Rev. 1) [12] can be consulted for this purpose.

The PSHA and/or DSHA may also include site response analysis; however, this is generally not the case. Most studies include site response analysis separately after the PSHA or DSHA results are obtained, for a 'bedrock' type condition. For this reason, this issue is addressed as part of Requirement 21 of SSR-1 [2].

Requirement 17 (see Section 4.2.6 of this Safety Report) and associated paras 5.6–5.10 of SSR-1 [2] address external hazard evaluation. Paragraphs 5.6–5.10 of SSR-1 [2] state:

"5.6. Capable volcanoes [footnote omitted] shall be identified and evaluated. The evaluation shall consider the volcanic characteristics of a region of sufficient size to ensure that potentially hazardous volcanic phenomena are considered appropriately.

"5.7. The hazards of capable volcanoes shall be evaluated to provide the input needed for determining the site specific design parameters or for re-evaluating the site, as well as for deterministic and/or probabilistic safety analyses performed during the lifetime of the nuclear installation.

"5.8. A proposed new site shall be considered unsuitable if reliable evidence shows the existence of a capable volcano that has the potential to affect the safety of the nuclear installation and which cannot be compensated for by means of a combination of measures for site protection and design features of the nuclear installation.

"5.9. An evaluation of volcanic hazards that focuses on determining the geological characteristics of volcanic phenomena and their spatial extent will usually be more certain than one focusing on an estimation of the likelihood of occurrence of hazardous phenomena. Volcanic hazards shall be evaluated using appropriate information, methods and models with adequate account taken of the uncertainties.

"5.10. The effect of volcanic phenomena in combination with other volcanically induced hazards shall be considered. This shall include consideration of volcanic ash fall."

Recommendations on volcanic hazards are provided in SSG-21 [11]; for example:

(a) Figure 1 of SSG-21 [11] divides the evaluation of volcanic hazards into four stages. The database requirements follow these stages (see Section 4 of SSG-21 [11]). The review also needs to follow these stages.
(b) Recommendations on the information needed for Stage 1 are provided in paras 4.4–4.8 of SSG-21 [11].
(c) For the detailed part of the work (if this is necessary as a result of Stage 1) — that is, for Stages 2, 3 and 4 — the database requirements are given in paras 4.9–4.37 of SSG-21 [11].
(d) Table 1 of SSG-21 lists 13 hazardous phenomena that may be associated with volcanoes. It also provides recommendations on whether these phenomena need to be considered exclusionary.

Data on some of these phenomena (e.g. tephra fallout) may need to be considered even if the volcano is hundreds of kilometres away. Other phenomena are much more local (e.g. volcano generated missiles, lava flows, ground deformation). In general, exclusionary criteria are associated with phenomena that are in close proximity to the site.

Volcanic hazard analysis may be deterministic and/or probabilistic. Volcano hazard analysis needs to consider at least the following parameters:

(a) Size of the volcanic event, which defines the distance at which different phenomena need to be considered;
(b) Characteristics of the volcano that may affect the extent of the phenomena (e.g. the viscosity of the lava flow is important in terms of the reach of the lava);
(c) Recurrence rate of volcanic phenomena.

When screening by probability is used, it is important that the screening probability level be chosen conservatively (e.g. generally at a level of 10^{-7}/year).

Significant aleatory and epistemic uncertainties may be associated with volcanic hazard modelling. If the screening of volcanic hazards is a contentious issue, consideration of the centre, body and range of technically defensible interpretations may also be useful in the volcanic hazard analysis.

If the volcanic hazard is associated with a monogenetic volcanic field, the method of analysis will differ from that used for volcanic hazards with prominent calderas. The sources and types of uncertainty may also be different. In cases of prominent calderas, it is important to estimate the locations and probabilities of the opening of new vents that may present different challenges.

Requirement 21 (see Section 4.2.6 of this Safety Report) and associated paras 5.24–5.26 of SSR-1 [2] address external hazard evaluation. Paragraphs 5.24–5.26 of SSR-1 [2] state:

"5.24. The static and dynamic geotechnical characteristics and geological features of subsurface materials at the site, including any backfill, shall be established. Laboratory and field-based methods shall be used, in conjunction with appropriate sampling techniques and sufficient repetition of each test, to characterize each parameter of the subsurface materials at the site.

"5.25. The stability and bearing capacity of foundation materials shall be assessed, including consideration of the potential for excessive settlement under static and seismic loading.

"5.26. The physical and the geochemical properties of the soil and groundwater shall be studied by appropriate methods and taken into account in the evaluation of the subsurface material at the site."

As mentioned above, site response analysis may be performed either as part of the PSHA or DSHA or as a separate task after these analyses are finalized. NS-G-3.6 [19] provides recommendations related to the necessary database and the methods used in site response analysis. Tables 1–3 of NS-G-3.6 [19] list the recommended methods for geophysical investigations, geotechnical investigations and laboratory testing. For generating the necessary database related to site response analysis, the most relevant methods are geophysical investigations coupled with the associated boreholes. According to table 1 of NS-G-3.6 [19], to obtain a reliable V_p and V_s profile down to approximately 100 m, the following geophysical investigations are needed:

(a) Seismic refraction or reflection;
(b) Cross-hole seismic test;
(c) Uphole or downhole seismic test;
(d) Nakamura method.

Depending on site conditions and availability, a sufficient combination of these methods needs to be used. To decrease the uncertainties related to these data, an effective QA programme needs to be in place, and the reviewer needs to confirm this.

Recommendations for input data and parameters for a site response analysis are provided in para. 3.7 of NS-G-3.6 [19], which emphasizes the importance of developing a detailed site model. This includes defining the geometry of soil layers, determining S and P wave velocities within each layer, assessing the soil's relative density, and using G–γ and η–γ curves, which represent how the shear modulus (G) and internal damping ratio (η) change with shear strain (γ) for each layer. For deep soil profiles, where wave velocities increase gradually with depth, the depth dependent variation of these parameters also needs to be taken into account.

To determine the appropriate method for calculating site response analysis and the uncertainties involved, paras 3.11 and 3.12 of NS-G-3.6 [19] may be used. Paragraphs 3.11 and 3.12 of NS-G-3.6 [19] state:

"3.11. In order to compute the site response, the following model is acceptable:

— A viscoelastic soil system overlying a viscoelastic half-space;
— A horizontally layered system;
— Materials that dissipate energy by internal damping;
— Vertically propagating body waves (shear and compression waves).

Non-linear effects may be approximated by equivalent linear methods. The equivalent linear model(s) of soil constitutive relationships should be consistent with the strain level induced in the soil profile by the response to the input ground motion. This generally leads to an iterative process.

"3.12. Uncertainties in the mechanical properties of the site materials should be taken into account through parametric studies, at least on the shear modulus value. One method is to vary the shear modulus between the best estimate value times $(1 + C_v)$ and the best estimate value divided by $(1 + C_v)$, where C_v is defined as the coefficient of variation. The minimum value of C_v is 0.5. Attention should be paid to the fact that a given soil profile cannot be assumed without an assessment to be conservative for all the items under consideration; that is, a conservative profile for deconvolution may not be conservative for the site response analysis."

As the ground motion prediction equations, or 'ground motion modelling', used in the seismic hazard analysis also include site characteristics (in terms of a V_{s30} parameter), it is important for the reviewer to check that double counting of site specific effects has not occurred.

Requirement 22 (see Section 4.2.6 of this Safety Report) and associated paras 5.27 and 5.28 of SSR-1 [2] address the evaluation of geotechnical and geological hazards. Paragraphs 5.27 and 5.28 of SSR-1 [2] state:

"5.27. The site and the site vicinity shall be evaluated to determine the potential for slope instability (such as landslides, rock fall and snow avalanches), caused by natural or human induced phenomena, which could affect the safety of the nuclear installation. In the evaluation of slope instability, the configuration of the site during and after site preparation activities shall be addressed. The evaluation of slope stability shall also take into account extreme meteorological conditions and rare meteorological events.

"5.28. The potential for slope instability resulting from seismic loading shall be evaluated using parameters appropriate for describing the seismic hazards and the soil and groundwater characteristics at the site."

Slopes near nuclear installations may be natural or artificial (i.e. built during site works). Once the design basis and the beyond design basis seismic levels have been determined, the stability of all slopes with the potential to endanger the structures, systems and components of the nuclear installation needs to be checked.

An acceptable safety margin for the beyond design basis seismic loads needs to be defined. A 10% margin for this load case is generally considered to be acceptable.

The data needed for slope stability analysis include the geometry of the slope, the history of slope failures, the material characteristics (e.g. cohesion, angle of internal friction), the water content and the seismic loading. If slope instability under seismic loads cannot be screened out, slope or soil improvement works may be performed. It is also possible to construct site protection features to avoid the landslide from advancing to the concerned structures, systems and components.

Uncertainties in measuring soil properties and other necessary data may be improved using effective quality control measures. These measures need to be checked by the reviewer.

Paragraph 5.29 of SSR-1 [2] states:

"The potential for collapse, subsidence or uplift of the surface that could affect the safety of the nuclear installation over its lifetime shall be evaluated using a detailed description of subsurface conditions obtained from reliable methods of investigation."

Collapse is a very local phenomenon that generally occurs in karstic terrain, where limestone formations are exposed to water intrusion. The consequence would be the creation of cavities in the rock. Karst can commonly occur in a variety of terrains; however, most of time the size and frequency of the cavities is manageable (i.e. they do not represent a safety concern to the nuclear installation). It is important, however, to investigate the situation very carefully and document the status of cavities down to a depth that has engineering significance. Table 1 of NS-G-3.6 [19] indicates two geophysical methods to be most effective for this purpose: microgravimetry and ground-penetrating radar. These geophysical investigations need to be used in conjunction with drilling to completely understand the depth and extent of the cavity and whether it is filled with soil. Uncertainties can be minimized through an effective QA programme. The civil engineering department needs to indicate whether the cavities can be tolerated by the foundation design or whether any remedial measures would be needed.

With respect to soil liquefaction, paras 5.30 and 5.31 of SSR-1 [2] state:

"5.30. The potential for liquefaction and non-linear effects of the subsurface materials at the site shall be evaluated using parameters appropriate for describing the seismic hazards and geotechnical properties of the subsurface materials at the site.

"5.31. The evaluation of soil liquefaction shall include the use of accepted methods for field and laboratory testing in combination with analytical methods to assess the hazards."

Soil liquefaction occurs in saturated granular soils and, if it is at a massive scale, it is considered an exclusionary attribute. Detailed recommendations on the investigations necessary to assess liquefaction potential, including groundwater regime, grain size distribution, standard penetration tests, cone penetration tests, relative density, and undrained cyclic strength, are provided in para. 3.16 of NS-G-3.6 [19]. The number of cycles of the vibratory ground motion is an important parameter in assessing liquefaction potential. Consequently, it is important that the analysis is performed in the time domain, with earthquake parameters compatible with the design basis and the beyond design basis seismic loading. Uncertainties can be minimized through an effective QA programme. The civil engineering department needs to indicate whether the potential settlements due to liquefaction can be tolerated by the foundation design or whether any remedial measures would be needed.

4.7.7. Reviews for small modular reactors

4.7.7.1. Vibratory ground motion hazard

Seismic hazard analysis is more advanced and widely used than other external hazard analyses because it cannot be excluded on the basis of site characteristics and because it needs ongoing updates to address regulatory changes and data uncertainties. Seismic hazard analysis relies on diverse geological and seismological data sets, which grow less reliable over time, necessitating detailed modelling of seismic sources and ground motion. These models help capture the variability and complexities of earthquakes. However, they also introduce epistemic uncertainties due to differences in expert interpretation, which may decrease as data expand and collaboration increases.

A formal methodology for integrating epistemic uncertainties within a PSHA was introduced by the Senior Seismic Hazard Analysis Committee (SSHAC) in the 1990s for a project involving all of the nuclear power plants operating in the United States. Since then, this method has been adopted and modified on the basis of experience in nuclear PSHA projects. SSG-9 (Rev. 1) [12] also refers to this approach in a footnote.

The SSHAC method has four levels of complexity and sophistication. For nuclear projects, there are examples of Level IV, Level III and Level II, as well as hybrid levels (e.g. Level II for seismic source modelling and Level III for ground motion characterization). The criteria for choosing an appropriate level for a PSHA are provided in Refs [22, 23]. Two examples of Level IV complexity under the SSHAC method are the Yucca Mountain project in the United States of America (for high level radioactive waste) and the Swiss Nuclear project PEGASOS, for all operating Swiss nuclear power plants. Some recent NUREG publications from the United States Nuclear Regulatory Commission describe SSHAC Level III in more detail; this definition has also found application in several European nuclear projects. One observation that was made relates to the long duration needed even for a SSHAC Level III. Part of the reason for this stems from the fact that the original SSHAC and PEGASOS were performed for operating nuclear power plants and therefore without much schedule pressure. However, this is not the case for new build projects.

One of the alternatives for simplifying the PSHA would be to use a lower level SSHAC process to include epistemic uncertainties. This alternative is worth exploring because it has a structured approach and ensures that the centre, body and range of technically defensible interpretations is adequately addressed.

To justify the use of simplified methods for SMRs, several important aspects of experience from large nuclear power plants need to be considered:

(a) There has generally been good experience for design basis exceedance of real earthquakes for nuclear power plants. Several nuclear power plants in Japan and the United States of America have experienced earthquakes in which the ground motion values exceeded the plants' design basis values. However, there have been no incidence of any structures, systems and components important to safety having failed because of this.

(b) No cliff edge effects such as flooding would be expected. Due to the failure mechanisms of structures, systems and components important to safety, it is understood that global seismic failure for a nuclear power plant is a progressive process and does not present a sharp cliff edge. This is an important attribute and one that has received much attention after the accident at the Fukushima Daiichi nuclear power plant.

(c) In SMRs, safety systems tend to be more passive than those in a large nuclear power plant. Although seismic motion will challenge these structures, systems and components, their failure would be progressive rather than sudden (compared with active components), thereby avoiding cliff edge effects.

Paragraph 9.10 of SSG-9 (Rev. 1) [12] provides specific recommendations related to the graded approach, focusing on the evaluation of hazards due to vibratory ground motion (see Section 3.3.1 of this Safety Report).

One of the safety targets for an SMR would be a categorization as described in item '(c)' in the list in Section 3.1.1 of this Safety Report. There are two alternatives in item '(c)', and here the choice for the PSHA would be for sub-item '(ii)', where the database and the methods can be reduced and simplified, but in return increased uncertainties need to be dealt with.

SMRs vary greatly in their design, and it cannot be automatically assumed that any SMR can be categorized in the same way using a graded approach. A graded approach needs to be applied step-by-step using the recommendations provided in SSG-9 (Rev. 1) [12] to determine the most appropriate categorization. Paragraph 9.1 of SSG-9 (Rev. 1) [12] states:

"The evaluation of seismic hazards for nuclear installations other than nuclear power plants should be commensurate with the complexity of such installations, with the potential radiological hazards and with the hazards due to other materials present on the site."

Increased uncertainties generally imply higher hazard values; therefore, any simplifications made in the methods need to be more conservative. In determining the application of a graded approach to seismic hazard evaluation for SMRs, the following should be considered, as applicable:

(a) Geological, geophysical and geotechnical database;
(b) Seismological database;
(c) Sophistication and complexity in seismic source modelling approaches;
(d) Sophistication and complexity in ground motion characterization approaches;
(e) Treatment of epistemic uncertainties;
(f) QA requirements;
(g) Type of independent review (e.g. participatory, end point).

The results of a graded approach adopted for an SMR will also be compared with the full SSG-9 (Rev. 1) [12] application. The main impact is expected to be associated with the following:

(a) Necessary human resources;
(b) Time needed for the conduct of PSHA;
(c) Resulting hazard curves.

A simplified approach might not be applicable to all types of seismotectonic setting, and some assumptions may need to be made regarding the site and the site region. Examples of these assumptions could include the following:

(a) The site may be affected both by faults and by area sources within the region.
(b) The site is not within the region of influence of a major subduction zone; instead, it is located within a region affected by shallow active crustal seismicity.
(c) The site is not within 100–150 km of a major transform fault.
(d) A reasonably well compiled historical and instrumental earthquake catalogue is available for the region.
(e) The site response characteristics can be managed through the use of appropriate ground motion modelling. The site condition is represented by a single V_{S30} value, and uncertainty in its measurement is disregarded for simplicity.
(f) The V_s-kappa adjustment to take into account crustal site amplification and high-frequency ground-motion attenuation is disregarded.
(g) Fault capability and its potential interaction with vibratory ground motion is outside the scope of the example.

(h)   Interactions between vibratory ground motion hazards and any seismically induced geotechnical hazard are ignored.

(i)   The structural frequency range may be narrower than for a large nuclear power plant and is therefore selected as appropriate for an SMR.

(j)   The annual frequency of exceedance will extend to 10^{-7}, as such low levels of annual frequency of exceedance may be needed for beyond design basis evaluation and seismic probabilistic safety assessment.

Due to the assumptions made above, and considering that the most resource intensive part of database acquisition is in the near region, it is not anticipated that the possible reduction in the regional geological, geophysical and geotechnical database will significantly impact cost and schedule. Simplification in the approaches to seismic source modelling and ground motion characterization will be the main focus in the simplification process.

4.7.7.2.   Probabilistic seismic hazard assessment methodology

The application of a graded approach to the PSHA methodology will primarily depend on the radiological risk associated with the SMR. However, other factors, such as unusual seismotectonic complexities related to a region, may also play a role in the selection of the SSHAC level to be used.

There may be a variety of ways to simplify the application of the recommendations provided in SSG-9 (Rev. 1) [12]. One such approach is provided in Ref. [16]. After verifying the necessary conditions, such an approach could be considered appropriate for an SMR.

4.7.7.3.   Fault displacement hazard

A fault displacement hazard is potentially exclusionary and, as recommended in SSG-9 (Rev. 1) [12], the application of a graded approach is not envisaged.

4.7.7.4.   Geotechnical hazards, including liquefaction, collapse, subsidence and slope instability (including those induced by seismic ground motion)

Geotechnical hazards are generally very localized and are potentially exclusionary in nature. However, it may be possible to improve the soil and/or rock conditions, and, depending on the extent to which these improvements are needed, the issue may become one of cost versus benefit.

One important point to consider for a multiple module site is the potential for geotechnical hazards to lead to common cause failures. For example, if there is a hill (either natural or human-made) near the site that might generate a landslide under seismic ground motion, which could damage multiple modules, very large safety margins would be needed for slope stability. Similarly, if multiple modules are founded on soil that might liquefy when subjected to high ground motion, the safety margins associated with this hazard need to be sufficiently high. These events are not only common failure causes, they can also lead to cliff edge effects.

In summary, the reviewer needs to verify that sufficiently large margins have been used in dealing with geotechnical hazards that may affect multiple modules. These margins would be higher than comparable margins used for a single large nuclear power plant.

4.7.7.5. *Volcanic hazards*

With regard to the exclusionary aspects of volcanic hazards (e.g. lava flows, pyroclastic flows, ground deformation), it is not appropriate to use a graded approach; therefore, the recommendations provided in SSG-21 [11] are fully applicable.

There may be some differences between an SMR and a large nuclear power plant in terms of screening distance values due to the possible difference in robustness to impact and blast loads. It is likely that larger screening distance values would be applicable to SMRs for these loads.

Otherwise, no application of a graded approach is envisaged for an SMR in relation to the evaluation of volcanic hazards; as such, the discussion in Section 4.7.6 of this Safety Report is applicable.

4.7.8. References for the review

The references for Section 4.7 include SSG-61 [1], SSR-1 [2], SSG-35 [10], SSG-21 [11], SSG-9 (Rev. 1) [12] and NS-G-3.6 [19].

4.8. SITE CHARACTERISTICS AND THE POTENTIAL EFFECTS OF THE NUCLEAR INSTALLATIONS IN THE REGION

4.8.1. Areas of review

The target of the SAR review in Section 4.8 includes the information recommended in para. 3.2.32 of SSG-61 [1].

Requirements 25–27 of SSR-1 [2] address the evaluation of the potential effects of the nuclear installation on the region in terms of the dispersion of radioactive materials, population distribution and public exposure, and uses of land and water. Section 4.8 mainly deals with Requirement 25 of SSR-1 [2], which states that "**The dispersion in air and water of radioactive material released from the nuclear installation in operational states and in accident conditions shall be assessed.**"

Section 4.8 interfaces with Sections 4.1, 4.5, 4.6 and 4.9–4.11 of this Safety Report.

The paragraph in SSG-61 [1] relevant to these areas of the review states:

"3.2.32. The characteristics of the site and the surrounding environment relevant to the dispersion of radioactive material in water, air and soil should be described in this section. The relevant requirements for evaluating the dispersion of radioactive material are established in Section 6 of SSR-1 [2]."

4.8.2. Acceptance criteria

Acceptance criteria for radiation protection are prescribed by national regulations. These criteria are given in terms of dose limits for workers and the public. Intermediate limits may also be imposed by the regulatory body, for example in terms of the concentration of various radionuclides in the air or in water.

To calculate the radioactive contamination in the air or in water, it is necessary to define the source term for discharges for both normal operation and accident conditions.

4.8.3. Review procedure

A typical safety review of an SAR consists of a desktop review of documents submitted by the applicant, site visits to confirm consistency between the SAR and the site condition, and independent calculations for verification.

The review is mainly performed from documents. Reviews already performed for the transport and diffusion of effluents in the atmosphere and hydrosphere may be revisited to determine the level of conservatism.

A site visit is necessary in order to understand the setting of the site and the contribution of the site characteristics with respect to the near region and site vicinity areas.

The review also needs to consider chapters 12 and 15 of the PSAR to put into context the different scenarios considered for postulated discharges from the nuclear installation.

4.8.4. Evaluation findings

The reviewer's evaluation of site characteristics, in accordance with the relevant regulatory criteria, needs to be documented in the SER, noting any findings that diverge from these criteria. The first step of the review is a decision on the overall quality and completeness of this Section of the SAR. It then needs to be verified that there are no issues related to site suitability.

The documentation of the review needs to include the evaluation of doses to workers and the public with respect to the relevant regulatory criteria. Interim values for the concentrations of different key radioisotopes, if applicable, need to also be compared against the regulatory criteria. The reviewer is responsible for independently verifying that the contents of the SAR comply with the acceptance criteria and for ensuring that the conclusions of the evaluation report also meet the acceptance criteria. The procedure followed in the evaluation needs to be briefly described. If needed, the evaluation may include the performance of independent calculations to ascertain compliance.

The SER needs to include short descriptions of the approaches adopted for the evaluation of the site characteristics and potential effects of the nuclear installations in the region, including the database used and the results obtained.

It is necessary to verify that the applicant has provided sufficient information and that the review and calculations (if applicable) support the conclusions of the SER.

4.8.5. Implementation

The requirements established in SSR-1 [2] and the recommendations provided in SSG-18 [13] and NS-G-3.2 [9] need to be used in the evaluations performed by the applicant, together with all applicable national regulations. If national safety targets and acceptance criteria related to doses to workers and the public are complied with using different approaches from those provided in the national regulations and IAEA safety standards, the methods proposed by the applicant need to be duly scrutinized.

4.8.6. Detailed review of technical content

Requirement 25 of SSR-1 [2] the dispersion of radioactive material (see Section 4.8.1 of this Safety Report). Associated requirements on atmospheric dispersion are established in paras 6.1 and 6.2 of SSR-1 [2] and are given in Section 4.6.6 of this Safety Report.

As mentioned in Section 4.6 of this Safety Report, part of the meteorological studies is related to the dispersion characteristics of radioactive effluents and for this purpose on-site measurements are generally needed throughout the lifetime of the nuclear installation. The release points of the nuclear installation are taken as the measurement points on a meteorological mast dedicated for this purpose.

Paragraph 6.1 of SSR-1 [2] indicates the type of data needed for modelling dispersion characteristics. These include wind speed, wind direction, air temperature, precipitation, humidity, atmospheric stability parameters and prolonged atmospheric inversions. One full year of locally obtained data (from the meteorological mast) is generally needed to address seasonal variations.

The measurements made locally involve sensitive instrumentation, and the reviewer needs to check that the maintenance and calibration of these instruments are conducted under a QA programme that ensures continuous and reliable data. The collection and processing of reliable data will have a major impact on reducing uncertainties. It is also expected that these measurements and data processing will be performed in accordance with written procedures.

Dispersion models developed using the meteorological data are used in the estimation of doses to workers and the public, including from a nuclear or radiological emergency at the nuclear installation. It is important that real time meteorological data are available to the emergency centre of the installation.

The requirements for radiological dispersion in bodies of water, established in paras 6.3–6.7 of SSR-1 [2], are discussed in Section 4.5.6 of this Safety Report. Additionally, the dispersion characteristics of radioactive effluents in surface water and groundwater are discussed in detail in Sections 4.5 and 4.7 of this Safety Report, respectively.

It is important to collect and eventually monitor the data related to surface water (e.g. rivers, lakes, oceans). These data include both physical and chemical characteristics.

The relationship between surface water and groundwater is an important topic, and a three dimensional hydrogeological model of the site vicinity needs to be constructed using the data from the monitoring programme. There will be uncertainties involved in the model, in particular related to parameters such as the transmissibility of the soil and rock matrix and the fractures and fissures within the matrix. The reviewer needs to check that these uncertainties have been appropriately considered in the model and that the results include a margin commensurate with these uncertainties.

Requirement 26 and associated paras 6.8–6.10 of SSR-1 [2] provide requirements on population distribution and public exposure (see Section 4.1.6 of this Safety Report).

Topics related to demography are discussed in Section 4.1 of this Safety Report. Paragraphs 6.8–6.10 of SSR-1 outline the data needed on resident and transient populations in the external zone of the nuclear installation. The data on population need to be displayed on maps, together with wind rose information obtained from the meteorological investigations.

The information related to resident population comes directly from census data. Information on the transient population may be more difficult to obtain from the census bureau, and it would be necessary to gather this data locally. Sites used by transient populations may include recreational facilities or schools, which may have seasonal variations. Special facilities such as military barracks and prisons need to be included in the local surveys.

Accounting for the present population will generally include some uncertainties; however, projecting population statistics to future years may involve substantial variability, depending on assumptions and projection models.

The reviewer needs to check that more than one model (e.g. optimistic, pessimistic, best estimate) has been used in the population projections and that the inherent uncertainties have been taken into account.

With regard to land and water in the region, Requirement 27 and associated para. 6.11 of SSR-1 [2] state:

"The uses of land and water shall be characterized in order to assess the potential effects of the nuclear installation on the region.

"6.11. The characterization of the uses of land and water shall include investigations of the land and surface water and groundwater resources that might be used by the population or that serve as a habitat for organisms in the food chain."

Land and water use in the external zone of a nuclear installation are important, especially in establishing the food chain and therefore the potential migration of radionuclides from the air, surface water or groundwater. The food chain is an important pathway in enabling accidentally released radionuclides to travel long distances. This could happen, for example, if food produced in the external zone has been contaminated and is transported to other parts of the country or even exported. Information on land and water use is especially important for the emergency plan for the nuclear installation. (See also Section 4.10 of this Safety Report.)

4.8.7. Reviews for small modular reactors

Currently, there is no explicit guidance on how to structure the external zone on the basis of specific SMR designs, population characteristics and meteorological conditions. For this reason, the reviewer needs to consider that a performance based approach is reasonable. Such an approach is recommended in

IAEA Safety Standards Series No. SSG-22 (Rev. 1), Use of a Graded Approach in the Application of the Safety Requirements for Research Reactors [24]. Paragraph 2.8 of SSG-22 (Rev. 1) [24] states:

"Qualitative categorization of the research reactor should be performed based on the potential radiological hazard, as follows:

(a) Facilities with significant potential for an off-site radiological hazard: such facilities include research reactors with high operating power, a large radioactive inventory or high pressure experimental devices. These facilities are categorized as a high potential hazard.

(b) Facilities with potential for an on-site radiological hazard only: such facilities include research reactors with an operating power up to a few megawatts, a limited radioactive inventory or with no high pressure experimental devices. These facilities are categorized as a medium potential hazard.

(c) Facilities with no potential radiological hazard beyond the research reactor hall and associated beam tubes or connected experimental facility areas: such facilities include facilities with low operating power, not requiring heat removal systems or with a small radioactive inventory. These facilities are categorized as a low potential hazard."

This concept can easily be adapted to SMRs. In fact, it is similar to the categorization suggested for the evaluation of external hazards for nuclear installations other than large nuclear power plants in Safety Guides such as SSG-9 (Rev. 1) [12], SSG-18 [13] and SSG-21 [11]. A brief review of the SMR designs in the IAEA SMR Book of 2024 [17] indicates that a categorization such as the one recommended in SSG-22 (Rev. 1) [24] may also be a useful tool for SMRs.

With regard to the categorization in para. 2.8 of SSG-22 (Rev. 1) [24]:

(a) Category (a) would include SMRs with a significant radioactive inventory, as well as SMR designs for which it cannot be demonstrated that the failure of a single module would not lead to multiple module involvement.

(b) Category (b) may include SMRs for which the source term and safety design demonstrate that, at most, only a single module could experience a severe accident (i.e. a single source term). Some micro-sized SMRs may also fall within this category.

(c) Category (c) would be reserved for micro-sized SMRs for which it can be demonstrated that the specified conditions are met. In addition, some SMR designs may claim that their design precludes the occurrence of a 'severe accident' and therefore qualify for Category (c).

In any case, the reviewer needs to verify that all of the claims are based on sound data, testing and analyses, with all uncertainties accounted for.

4.8.8. References for the review

The references for Section 4.8 include SSG-61 [1], SSR-1 [2], NS-G-3.2 [9], SSG-35 [10], SSG-18 [13] and SSG-22 (Rev. 1) [24].

4.9. RADIOLOGICAL CONDITIONS DUE TO EXTERNAL SOURCES

4.9.1. Areas of review

The target of the SAR review in Section 4.9 includes the information recommended in paras 3.2.33 and 3.2.34 of SSG-61 [1].

To understand the potential radiological impact of the nuclear installation on the region, background radiation needs to be measured before this impact starts. In general, measurements up to about 20 km from the installation are needed; monitoring needs to continue throughout the lifetime of the installation. To detect any changes in the background itself, some measurements beyond the 20 km radius are also needed.

Requirements 14 and 28 of SSR-1 [2] are the most relevant for these areas of the review.

Section 4.9 interfaces with Sections 4.8 and 4.11 of this Safety Report.

The paragraphs in SSG-61 [1] relevant to these areas of the review state:

"3.2.33. This Section should describe the radiological conditions in the environment at the site and in the surrounding area, with account taken of the radiological effects of other nuclear installations on the site and any other external radiation sources. The radiological conditions should be described in sufficient detail to serve as an initial reference point and a basis for future assessments of radiological conditions at the site and the surrounding environment.

"3.2.34. A description should be provided of the available radiation monitoring systems and the corresponding technical means for the detection of any radiation or radioactive contamination. If appropriate, this Section may reference other relevant sections of the safety analysis report concerned with the radiological aspects of licensing the plant."

4.9.2. Acceptance criteria

In general, there are no acceptance criteria associated with background radiation levels.

4.9.3. Review procedure

A typical safety review of an SAR consists of a desktop review of documents submitted by the applicant, site visits to confirm consistency between the SAR and the site condition, and independent calculations for verification.

The review is mainly performed from documents, which needs to include maps showing the locations of monitoring stations as well as any potential sources of radiation apart from the nuclear installation.

A site visit is necessary in order to check the information on the maps and the instrumentation for monitoring.

4.9.4. Evaluation findings

The reviewer's evaluation of site characteristics, in accordance with the relevant regulatory criteria, needs to be documented in the SER, noting any findings that diverge from these criteria. The first step of the review is a decision on the overall quality and completeness of this Section of the SAR.

The reviewer is responsible for independently verifying that the contents of the SAR comply with the acceptance criteria and for ensuring that the conclusions of the evaluation report also meet the acceptance criteria.

The SER needs to include short descriptions of the approaches adopted for the evaluation of radiological conditions due to external sources, including the database used and the results obtained.

It is necessary to verify that the applicant has provided sufficient information and that the review and calculations (if applicable) support the conclusions of the SER.

4.9.5. Implementation

The requirements established in SSR-1 [2] and the recommendations provided in SSG-61 [1] and NS-G-3.2 [9] need to be used in the evaluations performed by the applicant, together with all applicable national regulations. If different approaches from those provided in the national regulations and IAEA safety standards have been used, the methods proposed by the applicant need to be duly scrutinized.

4.9.6. Detailed review of technical content

With regard to data collection, Requirement 14 of SSR-1 [2] states:

"The data necessary to perform an assessment of natural and human induced external hazards and to assess both the impact of the environment on the safety of the nuclear installation and the impact of the nuclear installation on people and the environment shall be collected."

Associated requirements are established in paras 4.44–4.50 of SSR-1 [2]. In particular, paras 4.46 and 4.48 of SSR-1 [2] highlight site and regional environmental conditions, along with data for emergency response planning under various situations, and the need for periodic data review to reflect advances in data collection and analysis while ensuring relevance to evolving hazards.

To provide a framework for the results of the radiation monitoring of the nuclear installation site, radiation levels at the site and region (up to a 20 km radius) need to be determined. These will constitute the initial conditions for the radiation monitoring survey that will continue throughout the lifetime of the nuclear installation.

Requirements regarding the dispersion of radioactive material are established in Requirement 25 of SSR-1 [2] (see Section 4.8.6 of this Safety Report). Requirements regarding the monitoring of site conditions are established in Requirement 28 of SSR-1 [2] (see Section 4.11.6 of this Safety Report).

Requirements 12, 13, 44 and 45 of IAEA Safety Standards Series No. GSR Part 3, Radiation Protection and Safety of Radiation Sources: International Basic Safety Standards [25], can also be useful references in reviewing this Section of the SAR.

If the nuclear installation is being collocated together with already operating nuclear installations, the data from the monitoring network of the operation nuclear installation(s) may be used as the initial conditions for the new installation.

4.9.7. Reviews for small modular reactors

Practical review guidance provided in Section 4.9.6 of this Safety Report will generally apply but the suggested radius may be decreased using performance based criteria and in line with the categorization suggested in Section 4.8.7 of this Safety Report.

4.9.8. References for the review

The references for Section 4.9 include SSG-61 [1], SSR-1 [2], GSG-10 [8], NS-G-3.2 [9], SSG-35 [10] and GSR Part 3 [25].

4.10. SITE RELATED ISSUES IN EMERGENCY PREPAREDNESS AND RESPONSE AND ACCIDENT MANAGEMENT

4.10.1. Areas of review

The target of the SAR review in Section 10 includes the information recommended in paras 3.2.35–3.2.37 of SSG-61 [1].

Site characteristics might provide important inputs to the on-site and off-site emergency response plans. There are two major aspects to consider:

(a) Geographical and demographic characteristics of the site vicinity that may pose a hindrance to emergency plans. Sites located on islands or peninsulas may cause difficulties in moving populations during a nuclear emergency.
(b) Concurrent external hazards that may disable infrastructure planned for accident management or emergency response situations. Concurrent external hazards may have caused the emergency situation.

Requirement 13 of SSR-1 [2] and the recommendations in paras 6.1–6.8 of NS-G-3.2 [9] need to be considered. The requirements in IAEA Safety Standards Series No. GSR Part 7, Preparedness and Response for a Nuclear or Radiological Emergency [15], are also relevant.

Section 4.10 interfaces with Sections 4.1, 4.3 and 4.5–4.8 of this Safety Report.

The paragraphs in SSG-61 [1] relevant to these areas of the review state:

"3.2.35. The issues regarding feasibility of emergency preparedness in terms of plant accessibility and of transport of any equipment necessary in an emergency, including a severe accident, should be described in this section, with account taken of all reactor units and other nuclear and non-nuclear installations on the site, as applicable. The information provided should include the availability of adequate access and egress roads for evacuation of personnel, including access to and around the site, and supply networks in the vicinity of the site.

"3.2.36. The availability of local transport networks, communications networks and other infrastructure external to the site, during and after an external event, and issues regarding the feasibility of implementing emergency response actions should be described in this section.

"3.2.37. The need for any necessary administrative measures should be identified, together with the relevant roles of bodies and response organizations other than the operating organization."

4.10.2. Acceptance criteria

Depending on national regulations, acceptance criteria may include limitations for the siting of nuclear installations in certain areas.

4.10.3. Review procedure

A typical safety review of an SAR consists of a desktop review of documents submitted by the applicant, site visits to confirm consistency between the SAR and the site condition, and independent calculations for verification.

The review is mainly performed from documents, which can include maps showing the locations of on-site and off-site infrastructure to be used during the response to a nuclear or radiological emergency.

The review needs to focus on severe accident management plans and procedures to identify contingencies needed in case of a beyond design basis external hazard.

A site visit is necessary in order to check the information on the maps, the severe accident management procedures and the emergency plan.

4.10.4. Evaluation findings

The reviewer's evaluation of site characteristics, in accordance with the relevant regulatory criteria, needs to be documented in the SER, noting any findings that diverge from these criteria. The first step of the review is a decision on the overall quality and completeness of this Section of the SAR. It then needs to be verified that there are no issues related to site suitability.

The reviewer is responsible for independently verifying that the contents of the SAR comply with the acceptance criteria and for ensuring that the conclusions of the evaluation report also meet the acceptance criteria.

The SER needs to include short descriptions of the approaches adopted for the evaluation of site related issues in emergency preparedness and response and accident management, including the database used, the scenarios considered and the results obtained.

It is necessary to verify that the applicant has provided sufficient information and that the review and calculations (if applicable) support the conclusions of the SER.

4.10.5. Implementation

The requirements established in SSR-1 [2] and the recommendations provided in SSG-61 [1] and NS-G-3.2 [9] need to be used in the evaluations performed by the applicant, together with all applicable national regulations. If approaches different from those provided in the national regulations and IAEA safety standards have been used, the methods proposed by the applicant need to be duly scrutinized.

4.10.6. Detailed review of technical content

Requirement 13 (see Section 3.1 of this Safety Report) and associated paras 4.41–4.43 of SSR-1 [2] address the feasibility of emergency preparedness and response. Paragraphs 4.41–4.43 of SSR-1 [2] state:

"4.41. Requirement 13 applies also to the infrastructure of the external zone where emergency response actions might be warranted.

"4.42. An assessment shall be made of the feasibility of planning effective emergency response actions in accordance with GSR Part 7 [15]. Nuclear installations on the same site and at adjacent or nearby sites shall be considered in the assessment, with special emphasis on nuclear installations that could experience concurrent accidents.

"4.43. Any causal relationships between external events and the condition of the infrastructure on the site and in the external zone shall be considered when evaluating the feasibility of planning effective emergency response actions."

Demonstration of the feasibility of an effective emergency plan in relation to site conditions is a very important aspect of the site selection and site evaluation process. Three major potential impediments to effective emergency preparedness and response arrangements need to be considered, as follows:

(a) Geographical and/or topographical conditions of the site that may hinder evacuation planning. If, for example, a nuclear installation is located on or near an isthmus connecting a peninsula to the mainland, and a population centre is located at the distal end of that peninsula, the only land evacuation route would require an initial approach towards the source of the radioactive release.

(b) External hazards that may have played a role in the severe accident and that might also destroy infrastructure needed for sheltering and/or evacuation. It is also possible that some external hazards, even if they did not cause the accident, may have a sufficiently high probability of occurrence to be considered during an emergency response. Weather conditions such as sandstorms, dust storms or fog, and situations such as epidemics or pandemics are examples of these.

(c) Collocated nuclear installations that may also have simultaneous severe accidents due to a common cause.

Data, including data related to population distribution, access routes and weather conditions, need to be collected. The potential for traffic jams on access routes and for adverse weather conditions for sea travel need to be considered when response plans are prepared for nuclear or radiological emergencies.

If a nuclear or radiological emergency was caused by an external hazard such as an earthquake or flooding (or a combination of these), it is likely that this event was major and had the potential for massive devastation of the infrastructure in the near region of the site. Therefore, it is important to consider a variety of alternate ways in which sheltering and/or evacuation can be implemented. Such an event may also hinder on-site emergency protective actions and other response actions, and measures need to be taken so that severe accident procedures can be effectively implemented. One solution is to construct a very robust emergency response building within the nuclear installation site, at a location selected to remain functional even under beyond design external hazard conditions.

Collocated nuclear installations may be of different ages and built to different safety standards. Consequently, there is potential for an older installation to be in a severe accident situation and pose an 'external hazard' to the newer installation. This possibility needs to be considered, and necessary administrative measures need to be put in place.

The reviewer needs to verify that all of the data have been collected and that the lessons learned on this subject (e.g. Fukushima Daiichi nuclear power plant, Shoreham nuclear power plant) have been reviewed by the applicant in the evaluation of the feasibility of an effective emergency plan.

4.10.7. Reviews for small modular reactors

The feasibility of implementing effective emergency preparedness and response arrangements may be reviewed within the framework of the categorization suggested in Section 4.8.7 of this Safety Report. However, regardless of the size of external zone necessary, the three potential impediments to effective emergency preparedness and response arrangements described in Section 4.10.6 need to be taken into account, and transportation and communication within the site and with the outside community need to be ensured.

The three major potential impediments to effective emergency preparedness and response arrangements highlighted in Section 4.10.6 of this Safety Report also apply to SMRs. The reviewer needs to verify that all of these points have been investigated, even if the external zone for the SMR is reduced (or even within the site area of the SMR). Maintaining some communication and transportation links with the outside is important and needs to be ensured.

4.10.8. References for the review

The references for Section 4.10 include SSG-61 [1], SSR-1 [2], NS-G-3.2 [9] and SSG-35 [10].

4.11. MONITORING OF SITE RELATED PARAMETERS

4.11.1. Areas of review

The target of the SAR review in Section 4.11 includes the information recommended in paras 3.2.38–3.2.41 of SSG-61 [1].

Requirements 28 and 29 of SSR-1 [2] address the monitoring and review of external hazards and site conditions. NS-G-3.2 [9], SSG-21 [11], SSG-9 (Rev. 1) [12], SSG-18 [13], SSG-79 [18] and NS-G-3.6 [19] provide recommendations for monitoring site related parameters and reviewing site conditions.

Some of these activities require instrumental monitoring while others may simply involve updating information such as population statistics, infrastructure development and air corridors. Instrumentally monitored parameters may also be connected to alert systems that would link them to formal operating procedures. Flood and storm warnings are examples of these.

This Section 4.11 interfaces with Sections 4.1–4.10 of this Safety Report.

The paragraphs in SSG-61 [1] relevant to these areas of the review state:

"3.2.38. The strategy for monitoring site related parameters and the use of the results in preventing, mitigating and forecasting the effects of site related hazards should be described in this section.

"3.2.39. The provisions to monitor site related parameters affected by earthquakes and surface faulting, geological and volcanic phenomena, meteorological events, flooding, geotechnical hazards, hazards from biological organisms and human induced hazards (e.g. aircraft flight activities, chemical explosions, activities at nearby industrial and other facilities) should be described in this section. These provisions may be used for the following purposes:

(a) To provide the information necessary for operator actions taken in response to external events;
(b) To support the periodic safety review at the site;
(c) To develop models for the dispersion of radionuclides;
(d) To confirm the completeness of the set of site specific hazards taken into account.

"3.2.40. This Section should contain a description of the on-site meteorological monitoring programme, which can potentially be used for updating meteorological data in the future, for predicting the dispersion of radioactive substances during plant operation, or for early warning against extreme meteorological events. The monitoring of demographic and hydrological conditions over the lifetime of the plant should also be described in this Section (see SSR-1 [2]).

"3.2.41. Long term monitoring programmes should include the collection of data from site specific instrumentation and data from specialized institutions for use in comparisons to detect significant changes from the design basis, for example changes due to the possible effects of climate change."

4.11.2. Acceptance criteria

In general, acceptance criteria for data obtained from monitoring networks are indicated in QA documentation. For example, if a network of instruments is used in monitoring a site related parameter, there may be requirements that at least a certain percentage of the instruments are functional for the data to be accepted. Similar requirements may be valid for faulty or uncalibrated instruments.

Another criterion may be related to the duration of the monitoring. This is generally related to climate dependent data and necessitates that all seasons are covered in the monitoring period. Both meteorological and piezometric data to be collected for the PSAR need to be at least one year long to be considered acceptable.

4.11.3. Review procedure

A typical safety review of an SAR consists of a desktop review of documents submitted by the applicant, site visits to confirm consistency between the SAR and the site condition, and independent calculations for verification.

The review of monitoring programmes includes high level reviews such as of the objectives and appropriateness of the monitoring programme. However, detailed reviews are also needed. These are generally performed from QA documentation as well as field visits. The review needs to also consider how the results of the monitoring are used in improving the safety of the nuclear installation. For example, if the results from a monitoring programme show that the actual data are substantially different from the assumptions made for the analysis, corrective actions may be needed.

A site visit is necessary in order to check the locations of monitoring facilities and data processing stations at the projected nuclear installation.

An independent monitoring system needs to be installed and operated by the regulatory body at the nuclear installation site in addition to assessing data from the applicant's monitoring stations.

4.11.4. Evaluation findings

The reviewer's evaluation of site characteristics, in accordance with the relevant regulatory criteria, needs to be documented in the SER, noting any findings that diverge from these criteria. The first step of the review is a decision on the overall quality and completeness of this Section of the SAR. Next, it should be confirmed whether all of the acceptance criteria for the monitoring programmes have been complied with.

The reviewer is responsible for independently verifying that the contents of the SAR comply with the acceptance criteria and for ensuring that the conclusions of the evaluation report also meet the acceptance criteria.

The SER needs to include short descriptions of the monitoring approaches adopted, including the monitoring techniques used and the results obtained.

It is necessary to verify that the applicant has provided sufficient information and that the review and calculations (if applicable) support the conclusions of the SER.

4.11.5. Implementation

The requirements established in SSR-1 [2] and the recommendations provided in SSG-61 [1], NS-G-3.2 [9], SSG-21 [11], SSG-9 (Rev. 1) [12], SSG-18 [13], SSG-79 [18] and NS-G-3.6 [19] need to be used in the evaluations performed by the applicant, together with all applicable national regulations. If different approaches from those provided in the national regulations and IAEA safety standards have been used, the methods proposed by the applicant need to be duly scrutinized.

4.11.6. Detailed review of technical content

The topic of monitoring is related to many different variables and parameters. It covers all relevant environmental parameters as well as statistics on population, infrastructure development, and land and water use. The occurrence of external events is also monitored to check if they have an impact on the models that have been used in the corresponding hazard evaluation. Two requirements established in SSR-1 [2] address these topics, and they are treated separately in the rest of this Section of the Safety Report and need to be reviewed accordingly.

The main requirements regarding the monitoring of external hazards and site conditions are given in SSR-1 [2]. Requirement 28 and associated paras 7.1–7.3 of SSR-1 [2] state:

"All natural and human induced external hazards and site conditions that are relevant to the licensing and safe operation of the nuclear installation shall be monitored over the lifetime of the nuclear installation.

"7.1. The monitoring of external hazards and site conditions shall be commenced no later than the start of construction and shall be continued until decommissioning. The monitoring plan shall be developed as part of the objectives and scope of the site evaluation.

"7.2. The monitoring plan shall include the parameters to be monitored, the type of data to be collected, the methodology for data collection (including the location and frequency of data collection), the necessary resolution and precision of any measurements, data backup requirements, as well as requirements for data processing and analysis.

"7.3. Before commissioning of the nuclear installation begins, the levels of background radioactivity in the atmosphere, hydrosphere and lithosphere and in biota in the region shall be measured so as to make it possible to determine any additional radioactivity due to the operation of the nuclear installation."

The monitoring of all natural and human induced hazards and site conditions is performed in four different ways:

(a) Keeping track of and documenting all natural external events that occur in the region and checking that these events do not significantly impact the models and calculations that were used in assessing the hazard when the SAR was prepared (or revised). For example, if an earthquake occurs in the region, the seismotectonic model that was used in the seismic hazard analysis needs to be checked to ensure that it can accommodate such an event.

(b) Keeping track of and documenting all human induced events and potential changes in the sources of these events. Unlike most natural external events, human induced events are inherently non-stationary. Because of this, changes in the location, size and frequency of human induced events need to be monitored. It needs to be shown that these changes do not significantly impact the models and calculations that were used in assessing the hazard when the SAR was prepared (or revised).

(c) Instrumental monitoring of external events, including seismic monitoring and monitoring of volcanic activity. The former is generally routinely done (as recommended in SSG-9 (Rev. 1) [12]) and reported on a quarterly basis. The data are used in checking the validity of seismotectonic assumptions and in checking the activity of some tectonic features in the near region. Monitoring of volcanoes (when applicable) involves seismic monitoring networks as well as monitoring of groundwater characteristics.

(d) Instrumental monitoring of site characteristics, including radiological conditions. This type of monitoring includes a wide variety of site conditions, such as surface water (physical, chemical and biological characteristics), groundwater, foundation conditions (e.g. settlement) and meteorological conditions (from the on-site meteorological tower), as well as environmental monitoring.

The data related to (a) and (b) are generally obtained from official sources. If there is a possibility to do so, those data may be confirmed by the reviewer.

All instrumental monitoring needs to be conducted as part of the QA programme of the nuclear installation. It needs to be ensured that all instruments are well maintained and calibrated and that data processing is performed by qualified personnel using validated and verified software. This will reduce the uncertainties involved in the measurements. The reviewer needs to check that this is the case.

With regard to the review of external hazards and site conditions, Requirement 29 and associated paras 7.4 and 7.5 of SSR-1 [2] state:

"All natural and human induced external hazards and site conditions shall be periodically reviewed by the operating organization as part of the periodic safety review and as appropriate throughout the lifetime of the nuclear installation, with due account taken of operating experience and new safety related information.

"7.4. As part of periodic safety review (or as part of safety assessments conducted under alternative arrangements), natural and human induced external hazards and site conditions shall be reviewed throughout the lifetime of the nuclear installation using updated information. Such reviews shall be undertaken at regular intervals (typically no less than once in ten years), and in the event of any of the following:

(a) An update of the regulatory requirements;
(b) Indications of inadequate design against external hazards;
(c) New technical findings, such as the vulnerability of particular structures, systems and components to external hazards;
(d) New information, experience and lessons from the occurrence of actual external events that affected the safety of another nuclear installation or an industrial facility;
(e) Changes of hazards over time for which new information and assessments have become available;
(f) A need to provide additional confidence that there are sufficient margins to prevent cliff edge effects;
(g) As part of a programme for long term operation, or in support of an application for an extension to the operating licence for the nuclear installation;
(h) The development of new methods to analyse hazards that substantially improve earlier estimates.

"7.5. The site specific external hazards and the site conditions shall be re-evaluated, as necessary, based on the outcome of the periodic review of site specific hazards or because of new data relevant to the radiological environmental impact assessment or to the safe operation of the nuclear installation."

The monitoring of external hazards may lead to the re-evaluation of the external hazard. This may be because the recorded event has exceeded the design basis or has significantly impacted the calculations and models on which the hazard analysis was based.

There may be other reasons why there is a need to re-evaluate external hazards. These reasons are listed in para. 7.4 of SSR-1 [2]. The reviewer needs to check that all of these items have been reviewed and complied with by the applicant; the reviewer also needs to ascertain whether a hazard re-evaluation process has started or is being planned.

4.11.7. Reviews for small modular reactors

Different methods for monitoring are described in full in Section 4.11.6 of this Safety Report and are summarized here. The application of a graded approach to SMRs is discussed after each of these monitoring methods in the list that follows:

(a) Keeping track of and documenting all natural external events and checking that these do not significantly impact the models and calculations that have been used. The reviewer needs to check that this type of monitoring is being conducted and that a graded approach has not been applied. Natural external hazards such as large earthquakes and floods have the potential for common cause failures, and their occurrences need to be monitored very closely.
(b) Keeping track of and documenting all human induced events and potential changes in the sources of these events. The reviewer needs to check that this type of monitoring is also being conducted and that a graded approach has not been applied. Human induced external hazards also have the potential to exceed design bases and to affect multiple modules.
(c) Instrumental monitoring of external events, including seismic monitoring and monitoring of volcanic activity. Instrumental monitoring of volcanic activity needs to be maintained, and a graded approach is not appropriate if this hazard has been identified in the region. The need for routine seismic monitoring may be determined on a case-by-case basis, depending on the seismotectonic setting and the safety margins built into the SMR design. However, if the need for seismic monitoring is linked to the monitoring of a particular geological feature (e.g. a potentially capable fault, a volcano), the reviewer needs to verify the need for grading in accordance with the recommendations of SSG-21 [11] and SSG-9 (Rev. 1) [12].
(d) Instrumental monitoring of site characteristics, including radiological conditions.

The need for routine monitoring of site characteristics may be linked with the design of the SMR and the potential releases and discharges. Therefore, a performance based approach is possible. In addition to regulations specifically for nuclear installations, some types of environmental monitoring may be regulated by other national agencies, and these regulations need to be complied with.

4.11.8. References for the review

The references for Section 4.11 include SSG-61 [1], SSR-1 [2], NS-G-3.2 [9], SSG-35 [10], SSG-21 [11], SSG-9 (Rev. 1) [12], SSG-18 [13], SSG-79 [18], and NS-G-3.6 [19].

REFERENCES

[1]   INTERNATIONAL ATOMIC ENERGY AGENCY, Format and Content of the Safety Analysis Report for Nuclear Power Plants, IAEA Safety Standards Series No. SSG-61, IAEA, Vienna (2021).

[2]   INTERNATIONAL ATOMIC ENERGY AGENCY, Site Evaluation for Nuclear Installations, IAEA Safety Standards Series No. SSR-1, IAEA, Vienna (2019).

[3]   INTERNATIONAL ATOMIC ENERGY AGENCY, Governmental, Legal and Regulatory Framework for Safety, IAEA Safety Standards Series No. GSR Part 1 (Rev. 1), IAEA, Vienna (2016).

[4]   INTERNATIONAL ATOMIC ENERGY AGENCY, Licensing Process for Nuclear Installations, IAEA Safety Standards Series No. SSG-12, IAEA, Vienna (2010).

[5]   EUROPEAN ATOMIC ENERGY COMMUNITY, FOOD AND AGRICULTURE ORGANIZATION OF THE UNITED NATIONS, INTERNATIONAL ATOMIC ENERGY AGENCY, INTERNATIONAL LABOUR ORGANIZATION, INTERNATIONAL MARITIME ORGANIZATION, OECD NUCLEAR ENERGY AGENCY, PAN AMERICAN HEALTH ORGANIZATION, UNITED NATIONS ENVIRONMENT PROGRAMME, WORLD HEALTH ORGANIZATION, Fundamental Safety Principles, IAEA Safety Standards Series No. SF-1, IAEA, Vienna (2006),
https://doi.org/10.61092/iaea.hmxn-vw0a

[6]   INTERNATIONAL ATOMIC ENERGY AGENCY, Periodic Safety Review for Nuclear Power Plants, IAEA Safety Standards Series No. SSG-25, IAEA, Vienna (2013).

[7]   INTERNATIONAL ATOMIC ENERGY AGENCY, Functions and Processes of the Regulatory Body for Safety, IAEA Safety Standards Series No. GSG-13, IAEA, Vienna (2018).

[8]   INTERNATIONAL ATOMIC ENERGY AGENCY, UNITED NATIONS ENVIRONMENT PROGRAMME, Prospective Radiological Environmental Impact Assessment for Facilities and Activities, IAEA Safety Standards Series No. GSG-10, IAEA, Vienna (2018).

[9]   INTERNATIONAL ATOMIC ENERGY AGENCY, Dispersion of Radioactive Material in Air and Water and Consideration of Population Distribution in Site Evaluation for Nuclear Power Plants, IAEA Safety Standards Series No. NS-G-3.2, IAEA, Vienna (2002).

[10]   INTERNATIONAL ATOMIC ENERGY AGENCY, Site Survey and Site Selection for Nuclear Installations, IAEA Safety Standards Series No. SSG-35, IAEA, Vienna (2015).

[11]   INTERNATIONAL ATOMIC ENERGY AGENCY, Volcanic Hazards in Site Evaluation for Nuclear Installations, IAEA Safety Standards Series No. SSG-21, IAEA, Vienna (2012).

[12]   INTERNATIONAL ATOMIC ENERGY AGENCY, Seismic Hazards in Site Evaluation for Nuclear Installations, IAEA Safety Standards Series No. SSG-9 (Rev. 1), IAEA, Vienna (2022).

[13]   INTERNATIONAL ATOMIC ENERGY AGENCY, WORLD METEOROLOGICAL ORGANIZATION, Meteorological and Hydrological Hazards in Site Evaluation for Nuclear Installations, IAEA Safety Standards Series No. SSG-18, IAEA, Vienna (2011).

[14]   FOOD AND AGRICULTURE ORGANIZATION OF THE UNITED NATIONS, INTERNATIONAL ATOMIC ENERGY AGENCY, INTERNATIONAL LABOUR OFFICE, PAN AMERICAN HEALTH ORGANIZATION, WORLD HEALTH ORGANIZATION, Criteria for Use in Preparedness and Response for a Nuclear or Radiological Emergency, IAEA Safety Standards Series No. GSG-2, IAEA, Vienna (2011).

[15]   FOOD AND AGRICULTURE ORGANIZATION OF THE UNITED NATIONS, INTERNATIONAL ATOMIC ENERGY AGENCY, INTERNATIONAL CIVIL AVIATION ORGANIZATION, INTERNATIONAL LABOUR ORGANIZATION, INTERNATIONAL MARITIME ORGANIZATION, INTERPOL, OECD NUCLEAR ENERGY AGENCY, PAN AMERICAN HEALTH ORGANIZATION, PREPARATORY COMMISSION FOR THE COMPREHENSIVE NUCLEAR-TEST-BAN TREATY ORGANIZATION, UNITED NATIONS ENVIRONMENT PROGRAMME, UNITED NATIONS OFFICE FOR THE COORDINATION OF HUMANITARIAN AFFAIRS, WORLD HEALTH ORGANIZATION, WORLD METEOROLOGICAL ORGANIZATION, Preparedness and Response for a Nuclear or Radiological Emergency, IAEA Safety Standards Series No. GSR Part 7, IAEA, Vienna (2015), https://doi.org/10.61092/iaea.3dbe-055p

[16]   INTERNATIONAL ATOMIC ENERGY AGENCY, Optimization of Safety Measures for Protection of Nuclear Installations Against External Hazards, IAEA-TECDOC-2042, IAEA, Vienna (2024).

[17]   INTERNATIONAL ATOMIC ENERGY AGENCY, Advances in Small Modular Reactor Technology Developments: A Supplement to: IAEA Advanced Reactors Information System (ARIS) (2024 Edition), IAEA, Vienna (2024).

[18]   INTERNATIONAL ATOMIC ENERGY AGENCY, Hazards Associated with Human Induced External Events in Site Evaluation for Nuclear Installations, IAEA Safety Standards Series No. SSG-79, IAEA, Vienna (2023).

[19]   INTERNATIONAL ATOMIC ENERGY AGENCY, Geotechnical Aspects of Site Evaluation and Foundations for Nuclear Power Plants, IAEA Safety Standards Series No. NS-G-3.6, IAEA, Vienna (2004).

[20]   INTERNATIONAL ATOMIC ENERGY AGENCY, Design of Nuclear Installations Against External Events Excluding Earthquakes, IAEA Safety Standards Series No. SSG-68, IAEA, Vienna (2021).

[21]   INTERNATIONAL ATOMIC ENERGY AGENCY, The Fukushima Daiichi Accident, IAEA, Vienna (2015).

[22]   BUDNITZ, R.J., et al., Recommendations for Probabilistic Seismic Hazard Analysis: Guidance on Uncertainty and Use of Experts, Rep. NUREG/CR-6372, Lawrence Livermore National Laboratory, Livermore, CA (1997), https://doi.org/10.2172/479072

[23] NUCLEAR REGULATORY COMMISSION, Practical Implementation Guidelines for SSHAC Level 3 and 4 Hazard Studies, Rep. NUREG-2117 (Rev. 1), Office of Nuclear Regulatory Research, Washington, DC (2012).

[24] INTERNATIONAL ATOMIC ENERGY AGENCY, Use of a Graded Approach in the Application of the Safety Requirements for Research Reactors, IAEA Safety Standards Series No. SSG-22 (Rev. 1), IAEA, Vienna (2023).

[25] EUROPEAN COMMISSION, FOOD AND AGRICULTURE ORGANIZATION OF THE UNITED NATIONS, INTERNATIONAL ATOMIC ENERGY AGENCY, INTERNATIONAL LABOUR ORGANIZATION, OECD NUCLEAR ENERGY AGENCY, PAN AMERICAN HEALTH ORGANIZATION, UNITED NATIONS ENVIRONMENT PROGRAMME, WORLD HEALTH ORGANIZATION, Radiation Protection and Safety of Radiation Sources: International Basic Safety Standards, IAEA Safety Standards Series No. GSR Part 3, IAEA, Vienna (2014), https://doi.org/10.61092/iaea.u2pu-60vm

LIST OF ABBREVIATIONS

DSHA	deterministic seismic hazard assessment
EIA	environmental impact assessment
PSAR	preliminary safety analysis report
PSHA	probabilistic seismic hazard assessment
QA	quality assurance
SAR	safety analysis report
SER	safety evaluation report
SMR	small modular reactor
SSHAC	Senior Seismic Hazard Analysis Committee
UHS	ultimate heat sink

CONTRIBUTORS TO DRAFTING AND REVIEW

Al Senaani, H.S.	Federal Authority for Nuclear Regulation, United Arab Emirates
Aszodi, A.	Budapest University of Technology and Economics, Hungary
Contri, P.	International Atomic Energy Agency
Gurcharan, M.	Nuclear Regulatory Commission, United States of America
Gürpinar, A.	Consultant, Türkiye
Lee, H.	International Atomic Energy Agency
Lee, S.	Consultant, Republic of Korea
Sayin, B.	Nuclear Regulatory Authority, Türkiye
Spiler, J.	GEN Energija, Slovenia

Consultants Meeting

Vienna, Austria: 22–25 February 2022

Technical Meeting

Vienna, Austria: 9–13 October 2023

Feedback on IAEA publications may be given via the on-line form available at:
www.iaea.org/publications/feedback
The form may also be used to report safety issues or submit environmental queries concerning
IAEA publications.

Alternatively, contact IAEA Publishing directly:

Publishing Section, Dissemination Unit
International Atomic Energy Agency
Vienna International Centre, PO Box 100, 1400 Vienna, Austria
Telephone: +43 1 2600 22529 or 22530
Email: sales.publications@iaea.org
www.iaea.org/publications

ORDERING LOCALLY

To purchase priced IAEA publications, please contact either your preferred local supplier
or the IAEA's lead distributor:

Mare Nostrum Group

39 East Parade
Harrogate
North Yorkshire
HG1 5LQ
United Kingdom
Email: enquiries@mare-nostrum.co.uk
www.mngbookshop.co.uk

Trade Orders and Enquiries:

Telephone: +44 1243 843 291
Email: trade@wiley.com

Individual Orders and Enquiries:

Email: mng.csd@wiley.com

Priced and unpriced IAEA publications may also be ordered directly from the IAEA by contacting
IAEA Publishing. The recipient is responsible for shipping costs and any customs and duties.

Printed and bound by CPI Group (UK) Ltd, Croydon, CR0 4YY

06/07/2026

02160597-0005